CHEMICAL ENGINEERING METHODS AND TECHNOLOGY

ATOMIC LAYER DEPOSITION (ALD)

FUNDAMENTALS, CHARACTERISTICS AND INDUSTRIAL APPLICATIONS

CHEMICAL ENGINEERING METHODS AND TECHNOLOGY

Additional books in this series can be found on Nova's website under the Series tab.

Additional e-books in this series can be found on Nova's website under the e-book tab.

CHEMICAL ENGINEERING METHODS AND TECHNOLOGY

ATOMIC LAYER DEPOSITION (ALD)

FUNDAMENTALS, CHARACTERISTICS AND INDUSTRIAL APPLICATIONS

JEANNIE VALDEZ
EDITOR

nova publishers
New York

Library of Congress Cataloging-in-Publication Data

ISBN: 978-1-63483-869-6

Published by Nova Science Publishers, Inc. † New York

CONTENTS

PREFACE

Atomic layer deposition (ALD) is a thin film deposition technique used in the mass production of microelectronics. In this book, novel nonvolatile memory devices are discussed. The chapters examine the low-temperature fabrication process of single-crystal platinum non-thin films using plasma-enhanced atomic layer deposition (PEALD). A comprehensive review of ALD surface coatings for battery systems is provided, as well as a theoretical calculation on the mechanism of thermal and plasma-enhanced atomic layer deposition of SiO_2; and fluorine doping behavior in Zn-based conducting oxide film grown by ALD.

In: Atomic Layer Deposition (ALD)
Editor: Jeannie Valdez

ISBN: 978-1-63483-869-6

Chapter 1

FLUORINE DOPING BEHAVIOR IN ZN-BASED CONDUCTING OXIDE FILM GROWN BY ATOMIC LAYER DEPOSITION

Hyung-Ho Park, Young-June Choi and Kyung-Mun Kang

Department of Materials Science and Engineering, Yonsei University, Seoul, Korea

ABSTRACT

Recently, transparent conducting oxides (TCOs) as unusual materials have been attracted a lot of attention due to the recent development of semiconductor and display industry with a wide range of optoelectronic applications because they present high transparency with high conductivity, which are difficult to physically co-exist at the same time. ITO (Indium tin oxide; Sn-doped In_2O_3) is a well-known representative oxide of the TCO materials, but alternative material to replace ITO have been intensively investigated with the rise in the price of indium (In) due to the lack of its resource. For deposition of TCO films, spin-coating, sputtering, and chemical vapor deposition have been widely studied but, in recent years, atomic layer deposition (ALD) are emerging fast due to its advantages, including with the excellent quality of films without pin-hole, possibility of the deposition on the complex and micro size devices from its high step coverage. Therefore, atomic layer deposited ZnO (ALD-ZnO) could be a candidate material to alternate ITO because of its earth-abundance, non-toxicity, and good optical/electrical properties. Among recent investigations of suitable dopants for ZnO, most have used impurity-doped ZnO TCO films with metallic cation dopants, while there have been comparatively few studies of the TCO characteristics of ZnO doped with halogen anions. Furthermore, halogen anion-doped ZnO films have not been investigated using ALD due to the absence of halogen-related source. In this chapter, fluorine-doped ZnO films by ALD at low temperature were introduced through the fabrication by using home-made fluorine source. The fluorine doping concentration in the ZnO films according to the fluorine source pulse ratio were confirmed by using X-ray photoelectron spectroscopy and energy dispersive X-ray spectroscopy and, from this, the linearity of fluorine doping concentration in ZnO films could establish to control the fluorine composition in ZnO films. Furthermore, doping mechanism of fluorine anion in ZnO matrix was discussed with the growth mode of fluorine-doped ZnO films by analysis using X-ray diffraction, photoluminescence, and grazing-incident wide-angle X-ray diffraction. Finally, doping efficiencies of ALD-grown F-doped ZnO (ZnO:F) films could be calculated and as a result, the doping efficiency of ZnO:F films are three times higher than ZnO:Al films even though ZnO:F films were deposited at relatively low temperature. From these approaches, the doping behavior of fluorine as an extrinsic dopant for ALD-ZnO films can be established for optoelectronic applications. Moreover, the systematic studies were conducted in order to establish the effects of the properties of ZnO films doped with extrinsic dopants on the real optoelectronic device such as organic light-emitting diodes.

1. INTRODUCTION

According to a rapid increase of demands for use of smart display devices, transparent electronics become very important science and technology field, including materials science, chemistry, physics, and electrical/ electronic/ circuit engineering. In materials science, recently, the classes of materials combining good optical transparency with high electrical conductivity have grown rapidly to use for transparent electronics applications. New materials, such as graphene, carbon nanotube, and metal nanowire, have been challenged to apply them to transparent electrodes very recently [1-3]. This area, traditionally, was dominated by transparent conducting oxides (TCOs), which are oxide semiconductors with both electrical conductance and optical transparency at the same time, due to their wide application for transparent electrodes of next-generation optoelectronic applications in flat-panel displays (FPDs) [4], photovoltaic (PV) devices [5], and organic light-emitting diodes (OLEDs) [6]. Moreover, in response to the additional requirements of optoelectronic applications, TCO layers can be deposited on flexible, light-weight, and small plastic substrates at a temperature under the glass transition temperature of plastic substrates. In the material aspect, tin-doped indium oxide (indium tin oxide, Sn-doped In_2O_3; ITO) has been only commercially used for transparent electrodes of optoelectronic devices because of its excellent transparency in the visible region and high electrical conductivity. However, it is no longer new issue that the ITO should be substituted by indium-free TCO materials due to a number of problems of the use of ITO such as high cost due to the rarity of indium, expensive deposition techniques, environmental pollution, and the instability of indium in hydrogen plasma.

In order to replace ITO, other metal oxide semiconductors have been investigated very intensively. Among the alternative TCOs to ITO, zinc oxide (ZnO) and tin dioxide (SnO_2) is most widely investigated due to very low cost from high abundance in the earth, nontoxicity, and chemical stability in hydrogen plasma comparing with ITO. However, the electrical resistivity of intrinsic ZnO and SnO_2 (10^3~10^{-1} Ω cm) [7-11] is higher than that of ITO even though they have native n-type donors, such as oxygen vacancies and metal cation interstitials, as well as overlap of spherical 4s (ZnO) or 5s (SnO_2) orbitals consisting of the conduction band leads electrically conductive. Because ITO has high degeneracy from Sn doping in In_2O_3 matrix, representing high carrier concentration upto ~10^{21} cm^{-3} with relatively high mobility as high as ~35 cm^2 V^{-1} s^{-1}. Therefore, impurity doping in the ZnO or SnO_2 is essential requirement to improve electron carrier concentration in the films. Minami reported the published resistivity values of doped In_2O_3, ZnO, and SnO_2 reported from 1972 to 2005 in the review article [8]. The resistivity of impurity-doped ZnO films was linearly decreased as low as that of ITO through many efforts in order to optimize the electrical properties combining high electron concentration and high mobility by using various deposition techniques with various dopant materials [8, 12]. Thus, doped ZnO films as a transparent electrode, replacing ITO, will be realized in the near future with continued investigations.

The fabrication of TCO films has been investigated using several effective deposition techniques including chemical spray pyrolysis [13], chemical vapor deposition (CVD) [14], electron beam evaporation [15], pulsed laser deposition (PLD) [16], and radio frequency (rf) magnetron sputtering [17]. Recently, the atomic layer deposition (ALD) technique is growing up to fabricate TCO films because it has very powerful advantages for deposition of TCO films. ALD technique is a very advanced deposition method for low temperature-processing that yields highly uniform films. The ALD technique is a thin film growth method employing self-limiting surface chemistry comprising repeating processes of pulses and purges with precursors in which the source materials are kept separate during the entire deposition process. Therefore, ALD allows for low-temperature growth, better step coverage (high aspect ratio), and good uniformity, along with the ability to control film thickness through the number of ALD cycles. Moreover, the separate dosing of the precursors prevents gas phase reaction, which permits the use of highly reactive precursors and provides sufficient time for each reaction step to reach completion. These excellent properties allow for the deposition of electronic devices with complex 3D structures and for processing at relatively low temperatures, facilitating the use of flexible substrates. Moreover, in TCO research field, the new ideas and development method (work function control, surface texturing, and so on) have been investigated in order to overcome the limitation of material properties and apply in optoelectronic device, such as FPDs, PVs, OLEDs, and transparent thin film transistors (TFTs).

1-1. Transparent Conducting Oxide

In order to have high optical transparency, materials must have wide band gaps more than 3.3 eV to transmit the visible light spectrum (380 nm ~ 780 nm wavelength) and such a large band gap inhibits carrier doping in conduction band to be conductive. Therefore, it is very difficult that two physical properties, which are high optical transparency and high electrical conductivity, coexist in the same materials at the same time. For example, the electrical conductivity of metal films is very high more than 10^6 S cm^{-1} (lower than 10^{-6} Ω cm for resistivity) but they are not transparent. In order to be transparent metal films, the thickness must be reduced less than 10 nm. However, this very thin metal film is not chemically stable in air and has poor mechanical strength, so that the thin metal layer cannot be applied in commercial optoelectronic applications. In this point of view, TCOs are very exceptional materials because they have high optical transparency combined with high electrical conductivity. Since the discovery of cadmium oxide (CdO) by Badeker in 1907 [18], various binary metal oxides have been widely investigated as TCOs shown in Figure 1a [19]. Among these TCOs, the most common intrinsic TCOs are In_2O_3, ZnO, and SnO_2 and their band gaps are above that required for transparency across the full visible spectrum [20] as shown in Figure 1b.

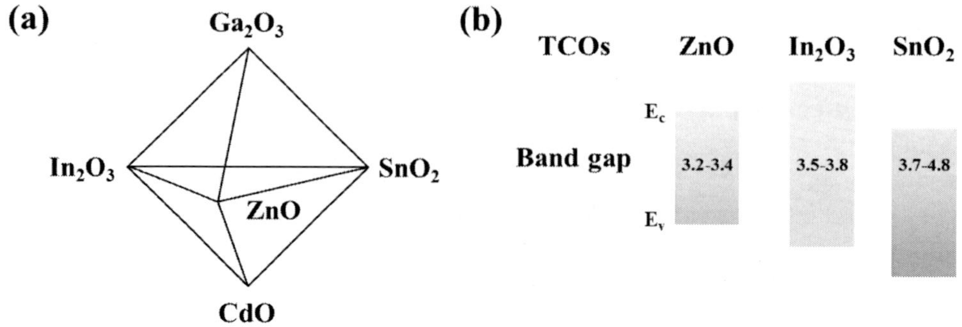

Figure 1. (a) Conventional TCO composition space and (b) band alignment of ZnO, In_2O_3, and SnO_2. E_C and E_V indicate the conduction band minimum and valence band maximum, respectively, and the numbers in the boxes are a range of optical band gap.

Commonly, intrinsic TCOs have mobilities in the range of ~10 $cm^2\ V^{-1}\ s^{-1}$ to 40 $cm^2\ V^{-1}\ s^{-1}$ and carrier concentration values in the range of ~$10^{18}\ cm^{-3}$ to ~$10^{19}\ cm^{-3}$ depending on the deposition techniques, post-treatment, and the types of substrates. As a result, resulted resistivity values are varied from ~$10^3\ \Omega$ cm to $10^{-3}\ \Omega$ cm higher than those of metals. Therefore, it is essentially necessary to dope with impurities in intrinsic TCO films in order to obtain high conductivity with high optical transparency because intrinsic TCO films can generate more carriers maintaining with large optical band gap when they are degenerately doped with impurities to displace the Fermi energy level up into the conduction band [21-23].

When the intrinsic TCOs are degenerately doped, the Fermi energy level clearly moves up above the conduction band minimum due to the Burstein-Moss (BM) effect [24, 25]. The shift makes to widen the optical band gap and to generate the carriers easily. Thus, many researches concerning with TCOs have been investigated the doping effects in TCO host in order to improve the electrical conductivity maintaining with high optical transparency. Recent years, ITO has been only commercially used among doped TCOs because of its excellent transparency and conductivity.

Table 1. Criteria for choosing TCOs reported by Gordon (Reproduced from [26])

Property application	Materials
Highest transparency	ZnO:F, Cd_2SnO_4, Zn_2SnO_4
Highest conductivity	In_2O_3:Sn
Highest mobility	CdO, In_2O_3:Ti, In_2O_3:Mo
Lowest plasma frequency	SnO_2:F, ZnO:F
Highest plasma frequency	In_2O_3:Sn
Lowest work function	ZnO:F
Highest work function	SnO_2:F, $ZnSnO_3$, ZnInSnO
Best thermal stability	SnO_2:F, Cd_2SnO_4
Best mechanical durability	SnO_2:F
Best chemical durability	SnO_2:F
Easiest to etch	ZnO:F
Lowest deposition temperature	In_2O_3:Sn, ZnO:B, a-InZnO
Least toxic	ZnO:F, SnO_2:F
Lowest cost	SnO_2:F

Table 1 shows the criteria for choosing TCOs reported by Gordon [26]. Zn presents the most abundant among cations in the aforementioned TCO materials, thus ZnO is the most cost-effective TCO materials so that the doped ZnO has been intensively explored because of its unique advantages.

1-2. Zinc Oxide (ZnO)

1-2-1. Structural Properties of ZnO

ZnO is a II-VI binary compound semiconductor, which mostly has crystal structure with either cubic zinc blende or hexagonal wurtzite phase where each cation is bonded by four anions at the corners of a tetrahedral coordination, and vice versa [27]. This coordination is typically composed with sp^3 hybridized covalent bonding nature, but also present a substantial ionic bonding character that leads to increase the band gap of these compound beyond the one expected from the covalent bonding. Thus, covalency or ionicity of ZnO resides at the borderline between the covalent and ionic semiconductors. The crystal structures of ZnO have rock salt, zincblende, and wurtzite as schematically shown in Figure 2 [27]. The Wurtzite ZnO is thermodynamically stable phase under ambient condition, while the zinc blend ZnO structure can be stabilized only by growth on cubic substrates and the rocksalt phase may be obtained at relatively high pressures [28].

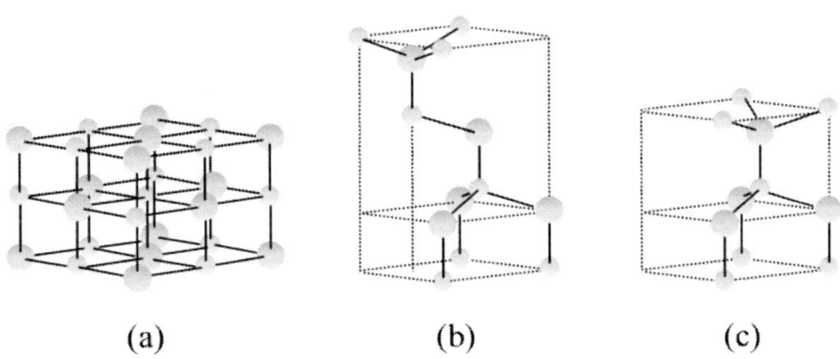

Figure 2. Three typical crystal structure of ZnO (a) rocksalt, (b) zincblende, and (c) wurtzite.

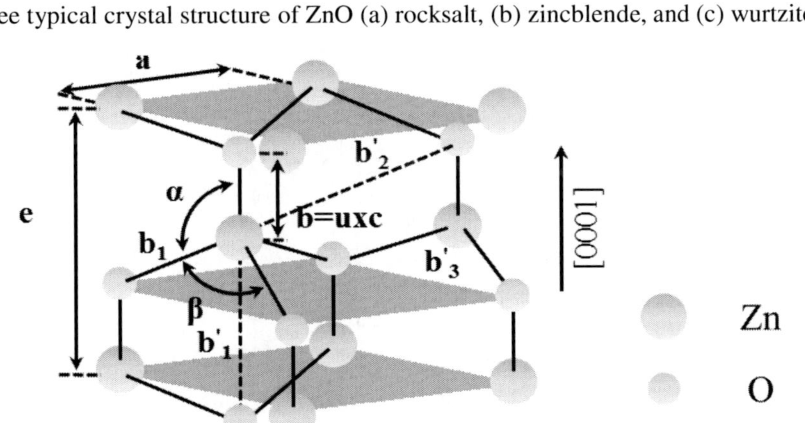

Figure 3. The schematic representation of the ZnO wurtzite structure.

Figure 3 shows a schematic representation of the ZnO wurtzite structure [27]. The wurtzite structure has a hexagonal unit cell with two lattice parameters *a* and *c* in the ratio of $c/a = \sqrt{8/3} = 1.633$ (in an ideal wurtzite structure) and belongs to the space group C_{6v}^{4} in the Schoenflies notation and $P6_3mc$ in the Hermann-Mauguin notation [27, 28]. The structure is composed of two interpenetrating hexagonal close-packed sublattices, each of which consists of one type of atom displaced with respect to each other along the threefold *c*-axis by the amount of $u = 3/8 = 0.375$ (in an ideal wurtzite structure) in fractional coordinates. The internal parameter *u* is defined as the length of the bond parallel to the *c*-axis (anion-cation bond length or the nearest-neighbor distance) divided by the *c* lattice parameter. The basal plane lattice parameter (the edge length of the basal plane hexagon) is universally depicted by *a*; the axial lattice parameter (unit cell height), perpendicular to the basal plane, is universally described by *c*. Each sublattice includes four atoms per unit cell, and every atom of one kind (group II atom) is surrounded by four atoms of the other kind (group IV), or vice versa, which are coordinated at the edges of a tetrahedron [27, 28].

1-2-2. Optical Properties of ZnO

In TCOs, the three main regions of transmittance are typically divided into ultra-violet (UV), visible, and near-infrared (NIR). First, as the carrier concentration of the films increased through impurities doping in ZnO, the absorption was increased in the NIR region. This phenomenon can be explained by the classical Drude free-electron theory. In other words, absorption by free electrons restricts the transmittance of films with high electron density in the conduction band such that the absorption and reflection of electron gas (plasmon) increases and the transmittance decreases for frequencies lower than the plasma frequency (for wavelengths higher than the plasma wavelength) as follows:

$$\omega_P^2 = \frac{ne^2}{\varepsilon_{ZnO}\varepsilon_0 m_e^*} \qquad \text{(Eq. 1)}$$

$$\lambda_P = \frac{c}{\omega_P} \qquad \text{(Eq. 2)}$$

where ω_P is the plasma frequency; *n* is the electron concentration; *e* is the electron charge; ε_{Zno} and ε_0 are the dielectric constant measured in the transparent region of the spectrum of undoped ZnO and the vacuum permittivity; λ_P is the plasma wavelength; and *c* is the velocity of light in a vacuum [29]. Therefore, with an increase in the carrier concentration with an increase in the impurity doping concentration, the absorption of the films was increased by the shortening of the plasma wavelength. These results are related with the electrical properties of undoped and doped ZnO. In the literature [26], the plasma wavelengths of ZnO:F and ZnO:Al films were reported to be greater than 2000 nm and 1300 nm, respectively. Generally, anion (halogen)-doped ZnO films compared with cation-doped ZnO films have higher plasma wavelength; namely, the higher mobility of the halogen-doped ZnO films led to low resistivity even though the carrier concentration was lower than in cation-doped ZnO films. This yields films that are suitable for applications in solar cells for long wavelength light absorption. In visible region, undoped ZnO generally shows high transmittance over 80% with glass substrate because of its wide band gap. In the case of doped ZnO films, the optical band gap of ZnO films is rather increased with an increase of carrier concentration till a certain point due to band gap widening from BM effect. However, with an increase of carrier concentration, the band gap narrowing (band gap renormalization) is able to occur due to many-body interaction shown in Figure 4 [30, 31]. Figure 4 shows schematic band structure diagram for TCO film with an increase of degeneracy [30].

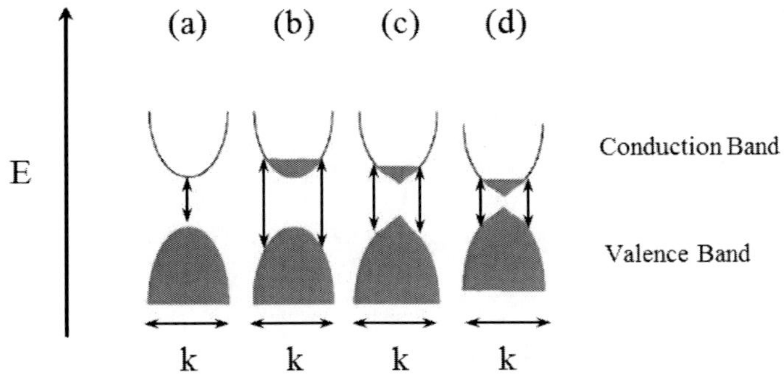

Figure 4. Schematic band structure diagram for representative TCO, In_2O_3. (a) Typical band structure of direct band gap oxide, (b) band gap widening with the BM shift, (c) and (d) band gap narrowing due to band gap renormalization from many-body interactions.

To further discuss the optical behaviors of ZnO in the UV region, from transmittance data, the absorption coefficient was calculated using Lambert's formula:

$$\alpha = -\frac{1}{t}\ln T_r \qquad \text{(Eq. 3)}$$

where t and T_r are the film thickness and transmittance, respectively [32]. The direct band gaps of the films were estimated from the $(\alpha h\nu)^2$ vs. $h\nu$ (Tauc relation) plot by extrapolating the fit of the linear region to $\alpha = 0$. The estimated optical band gap values give information about relationship between the optical behaviour and electronic structure of ZnO films. The shift in the absorption edge increased with increasing amounts of carrier concentration. The increase in the band gap with respect to increasing carrier concentration is generally believed to be due to the BM effect [24, 25].

ZnO films are n-type oxide semiconductors due to the electron donors generated by O vacancies and Zn interstitials. The addition of dopant raises the Fermi level of the doped ZnO films into the conduction band, causing the films to completely degenerate, and the absorption edge therefore moves to an energy higher than the actual band gap of pure ZnO. Finally, the increase in the absorption in the NIR region and the optical band gap in the UV region could be considered to be additional evidence of the increase in electron concentration by impurity doping. The amount of the BM shift can be calculated by Eq. 4 [33, 34]. Assuming parabolic energy band, and due to direct-band gap of ZnO, the optical band gap widening can be presented by the energy [35],

$$\Delta E_{BM} = \left(\frac{h^2}{8m^*}\right)\left(\frac{3}{\pi}\right)^{2/3} n^{2/3} \qquad \text{(Eq. 4)}$$

where h is Plank's constant, m^* is the reduced effective mass, and n is the electron concentration in ZnO or doped ZnO films.

1-2-3. Electrical Properties of ZnO with Defects and Extrinsic Doping

The conduction band of ZnO arises from the empty 4s levels of Zn^{2+} (stressing the ionic binding) or from the lowest anti-bonding sp^3 hybrid states (stressing covalent binding). In the same sense, the valence band of ZnO originates from the occupied 2p states of O^{2-} or the energetically highest bonding sp^3 states [36]. Therefore, undoped ZnO is always n-type due to native n-type donors (native defects in ZnO), such as O vacancies (V_O) and Zn cation interstitials (I_{Zn}) [37], as well as overlap of spherical 4s (ZnO) orbitals consisting of the conduction band leads electrically conductive. In intrinsic ZnO, one O vacancy or Zn interstitial is always accompanied by two electrons to maintain the charge balance as follows.

$$O_O^x = \frac{1}{2}O_2(g) + V_O^{\bullet\bullet} + 2e' \qquad \text{(Eq. 5)}$$

$$Zn_0^x = Zn_i^{\bullet\bullet} + 2e' \qquad \text{(Eq. 6)}$$

However, conductivity of intrinsic ZnO is not sufficient to apply in optoelectronic applications due to limitation of carrier concentration from native donors. Therefore, extrinsic doping in ZnO is essentially needed. The most widely used dopants are group III elements, such as Al, Ga, and B. In the case of metal dopant (group III element), all dopant atoms (substitution to Zn) make a strong local perturbation to the conduction band, which causes strong scattering of the electrons in the conduction band with relatively high carrier concentration (over 10^{21} cm^{-3}), thereby reducing their mobility. While halogen dopant (group group VII: F, Cl), which substitute for O, would perturb mainly the valence band, so that the conduction band is relatively free of scattering and it leads relatively higher electron mobility. Due to higher mobility, halogen dopant acts as an effective n-type dopant with relatively small amount of dopant concentration in ZnO matrix, which represents low carrier concentration in ZnO films (under 10^{21} cm^{-3}).

1-3. Deposition Techniques for ZnO Thin Films

1-3-1. RF Magnetron Sputtering

ZnO films are deposited at a certain growth temperature with a ZnO target material by using rf magnetron sputtering. The deposition is generally performed in an atmosphere with O_2 or Ar/O_2 mixed gas at a pressure ranging from 10^{-3} to 10^{-2} Torr [38]. O_2 and Ar gases play a role as the reactive gas and the sputtering enhancing gas. RF magnetron sputtering is most widely used to grow ZnO thin films, because it can assure low cost of target materials, simple processing, and low operating temperature [39]. Post treatment such as post-annealing, however, is surely used to relieve the physical stress of film surface and enhance the properties of ZnO films [40, 41].

1-3-2. Molecular Beam Epitaxy (MBE)

MBE has the advantage of accurate controllability of the film growth parameters because reflection high-energy electron diffraction provides a real time monitor for the growth mode of epitaxial ZnO layers [42]. MBE-deposited ZnO thin film typically use high-purity Zn metal source and O_2 radical beam. MBE technique is good for epitaxial growth of ZnO layer but growth rate and thickness uniformity is very poor comparing with other ZnO deposition methods. And also, high vacuum conditions have to be required for vaporization and transport of source materials to substrate. Thus, it has a limitation for a commercial application of ZnO films.

1-3-3. Pulsed Laser Deposition (PLD)

Compared to other ZnO deposition techniques, PLD is a relatively simple deposition technique and demands a minimum level of optimization to establish thin film growth of material composition [43]. In the PLD technique, oxidation occurs primarily in the ZnO ablation plume and this alleviates the difficulties encountered with other techniques where oxidation proceeds via surface reactions. Thin film is obtained by vaporizing a material using high-energy laser pulses. The target composition is transferred instantaneously leading to a stoichiometric deposition. It is generally easier to obtain the desired stoichiometry in the deposited films than other deposition technologies. For ZnO thin film deposition, a variety of target materials including sintered ceramic disc prepared from pressed power [44], and single crystals of ZnO [45] have been reported, and sintered ceramic targets are most commonly used. This technique is ideally suitable for deposition on small substrates, and yields high quality film useful for research and allows fabrication of discrete devices. However, scaling for industrial production and deposition on large area substrates is still the main practical limitation [46].

1-3-4. Chemical Vapor Deposition (CVD)

CVD is interesting because it is a scalable process making it possible to grow uniform large-area films. ZnO deposition occurs as a result of chemical reactions of vapor-phase precursors on the substrate, which are delivered into the growth zone by the carrier gas. The reactions take place in a reactor where a necessary temperature profile is created in the gas flow direction. For ZnO growth, CVD technique typically involves the use of metal alkyls, usually dimethyl zinc [$(CH_3)_2Zn$] (DMZ) or diethyl zinc [$(C_2H_5)_2Zn$] (DEZ) in combination with a separate source of oxygen and argon or nitrogen as a carrier gas. O_2 or H_2O are typical oxygen precursors in CVD growth of ZnO films [47, 48]. However, DEZ and DMZ are highly reactive with oxygen and water vapor resulting in a premature gas phase reaction in the cold zone of the reactor. This premature reaction results in the formation of ZnO powder, which degrades the film quality. Thus, high-quality ZnO has been formed by CVD using improved reactor design and the use of less-reactive precursors. Stable metal-organic source of zinc acetylacetonate in combination with oxygen was successfully used for the growth of high-quality ZnO films on *r*-plane [49] as well as on *c*- and *a*-plane [50] sapphire substrates by atmospheric pressure MOCVD. A variety of oxygen compounds were employed in MOCVD such as isopropanol (*i*-PrOH) [51], tertiary-butanol (*t*-BuOH) [52], acetone [53], N_2O [54], NO_2 [55], and CO_2 [56]. Plasma enhanced chemical vapor deposition (PECVD) technique was also used for high quality ZnO deposition [56]. It has several advantages in preparing ZnO thin films. First, a film densification process induced by plasma bombardment and ultraviolet ray irradiation can improve the thin-film quality [56]. This low-energy plasma can also enhance the surface reactions. Second, since the reactants used in PECVD are decomposed by the plasma but not by thermal energy, the growth does not need a high temperature environment [57] and makes it possible for ZnO thin-film to be prepared on the flexible substrate and on the organic material.

1-3-5. Atomic Layer Deposition (ALD)

Recently, ALD has been actively investigated to grow ZnO thin films at a relatively low temperature. The ALD technique is a very advanced deposition method for low temperature-processing that yields highly uniform films. The ALD technique is a thin film growth method employing self-limiting surface chemistry comprising repeating processes of pulses and purges with precursors in which the source materials are kept separate during the entire deposition process. Therefore, ALD allows for low-temperature growth, better step coverage (high aspect ratio), and good uniformity, along with the ability to control film thickness through the number of ALD cycles. Moreover, the separate dosing of the precursors prevents gas phase reactions, which permits the use of highly reactive precursors and provides sufficient time for each reaction step to reach completion. These excellent properties allow for the deposition of electronic devices with complex 3D structures and for processing at relatively low temperatures, facilitating the use of flexible substrates.

1-4. Atomic Layer Deposition

1-4-1. Definition of ALD

ALD is a CVD method suitable for fabricating inorganic material layers with thickness down to a fraction of a monolayer [58]. ALD has the capability to coat extremely complex shapes with a conformal material layer of high quality, a capability unique among thin-film deposition techniques [58, 59]. Consequently, ALD-grown materials have a wide range of applications, from catalysts to electroluminescent displays to microelectronics and beyond [58, 59]. ALD can be defined as a film deposition technique that is based on the sequential use of self-terminating gas-solid reactions [60, 61]. From this definition, various materials can be fabricated by ALD [62].

1-4-2. Principle of ALD

The growth of material layers by ALD consists of repeating the following characteristic four steps:

(i) A self-terminating reaction of the first reactant (Reactant A).
(ii) A purge or evacuation to remove the non-reacted reactants and the gaseous reaction by-products.
(iii) A self-terminating reaction of the second reactant (Reactant B) or another treatment to activate the surface again for the reaction of the first reactant.
(iv) A purge or evacuation.

Steps (i)~(iv) constitute a *reaction cycle*. Steps (i) and (iii) are sometimes referred to as *half reactions* of an ALD reaction cycle [63]. Each reaction cycle adds a given amount of material to the surface, referred to as the growth per cycle (GPC). To grow a material layer, reaction cycles are repeated until the desired amount of material has been deposited [64].

Before starting the ALD process, the surface is stabilized to a known, controlled state, for example, by a heat treatment. The use of self-terminating reactions leads to the conclusion that ALD is a surface-controlled process, where process parameters other than the reactants, substrate, and temperature have little or no influence. Because of the surface control, ALD-grown films are extremely conformal and uniform in thickness.

2. F-DOPED ZNO FILMS BY ATOMIC LAYER DEPOSITION

2-1. Novel Approach to the Fabrication of F-Doped ZnO Films by Atomic Layer Deposition at Low Temperatures

A simple low-temperature fabrication of fluorine-doped ZnO (ZnO:F) thin films by ALD was investigated and the growth mode of the films was analyzed. A novel method for fluorine doping in a ZnO matrix was successfully developed that uses ALD with home-made fluorine source at a low deposition temperature of 140 °C, which is low enough for films to be applied to a plastic substrate. The fluorine doping concentration was controlled from 0 at.% to 1.2 at.% by manipulating the pulse sequence ratio of the oxygen source (DI water) and the home-made fluorine source. Structural and morphological properties were investigated by ray diffraction (XRD) and field emission scanning electron microscopy (FE-SEM). The growth mode of the films was studied by grazing-incidence wide-angle XRD. The grain growth orientation was found to change significantly as the fluorine concentration was increased due to the characteristics of fluorine doping in the oxygen sites of ZnO. This phenomenon could be explained by a perturbation and passivation effect resulting from the fluorine doping mechanism, with the fluorine anions filling oxygen-related defect sites in the ZnO lattice. A photoluminescence study confirmed the diminishment of defect sites and showed that, due to the BM effect, the optical transmittance in the visible region increased significantly with an increase in the fluorine doping concentration from 80.87% to 83.02%. The lowest resistivity was 1.876×10^{-3} Ω cm for ZnO thin films doped with 1.0 at.% fluorine when the carrier concentration and mobility were 1.375×10^{20} cm^{-3} and 24.20 cm^2 V^{-1} s^{-1}, respectively.

2-1-1. Experimental Details

2-1-1-1. Novel Atomic Layer Deposition of ZnO:F Films at Low Temperatures

Undoped ZnO and ZnO:F thin films were deposited on thermally oxidized SiO_2 (100 nm) on Si substrates, Si substrates, and LCD glass (Fusion 1737) substrates via ALD using a traveling wave type Lucida D100 system (NCD Technology, Inc. Korea) at a deposition temperature of 140 °C and a working pressure of less than 1 Torr. DEZ and DI water were used as a precursor for Zn and an O reactant, respectively. A home-made F source was created by mixing 50 mL DI water and 0.5 mL diluted hydrogen fluoride (HF 48~51% diluted in water). This solution was stirred for 5 min in order to uniformly disperse the HF in DI water before the solution was charged to a canister. The chemical state of this source solution is composed of hydrofluoric acid and/or liquid hydrogen fluoride (Figure 5). DEZ was kept below 10 °C by using a chiller and delivered into a chamber with a high purity N_2 (99.999%) carrier gas flow of 20 sccm (standard cubic centimeters per minute). ALD growth cycles for undoped ZnO thin films were as follows: DEZ pulse (0.1 s) → N_2 purge (10 s) → H_2O pulse(0.1 s) → N_2 purge (10 s) (Figure 6a). In order to incorporate the desired F doping content into ZnO thin films, water (H_2O) and the home-made F source (H_2O/HF) were alternately pulsed with a controlled x:y ratio. For example, for Zn-O:Zn-O/F (4:1) samples, one super cycle comprised four ($x = 4$) repeats of the ALD sequence, DEZ (0.1 s) pulse → N_2 purge (10 s) → H_2O pulse(0.1 s) → N_2 purge (10 s) and one ($y = 1$) ALD sequence, DEZ (0.1 s) pulse → N_2 purge (10 s) → H_2O/HF pulse (0.1 s) → N_2 purge (10 s) (Figure 6b). The sequence of the ALD-ZnO:F films with single reactant (H_2O/HF) was DEZ pulse (0.1 s) → N_2 purge (10 s) → H_2O/HF pulse (0.1 s) → N_2 purge (10 s), and the F concentration could be maximized from this ALD sequence (Figure 6c).

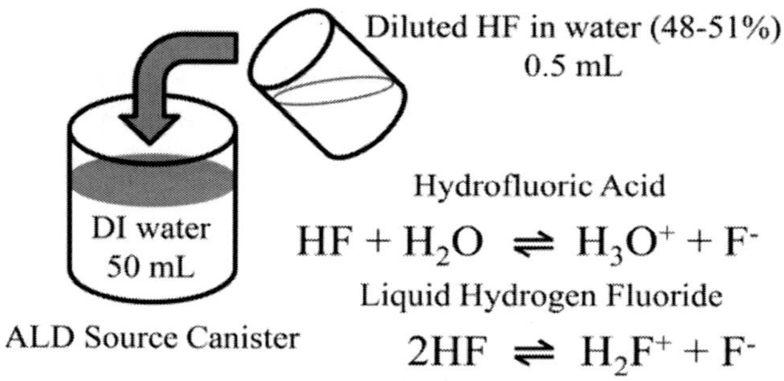

Figure 5. Illustration of the fabrication method for the home-made fluorine source and the predicted chemical state of the fluorine source solution [65].

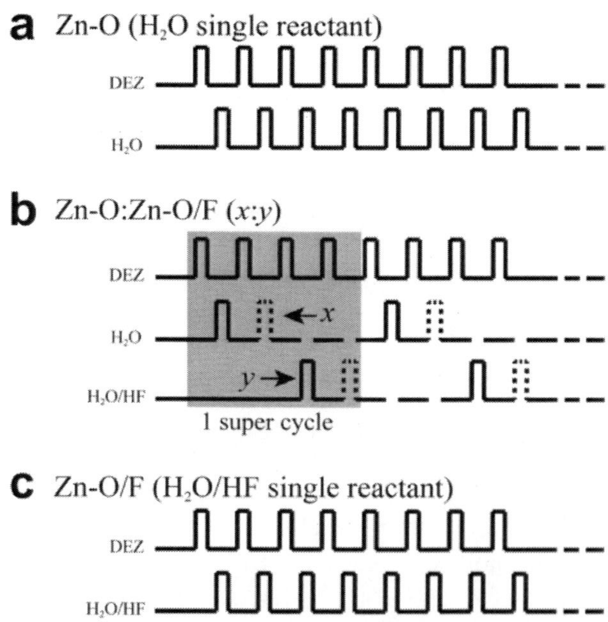

Figure 6. Schematic diagram of the process sequence used to grow (a) undoped and (b,c) fluorine-doped ZnO thin films. O/F indicates the fluorine single source of mixed H_2O and diluted HF (48~51% in H_2O). In (b), x is the pulse ratio of the H_2O source and y is the pulse ratio of the fluorine source (H_2O/HF) in one super cycle. The source pulse time and high purity N_2 purge time were fixed at 0.1 and 10 s, respectively [126].

2-1-1-2. Extrinsic F Doping in ZnO Films via ALD

Undoped ZnO thin films were fabricated using the ALD process sequence (Figure 6a). The general ALD reaction for the fabrication of ZnO films is very simple and has been reported [66, 67]. The overall reaction during the formation of ZnO films can be generally defined by formulas (Eq. 7) and (Eq. 8):

$ZnOH^* + Zn(C_2H_5)_2 \rightarrow ZnOZnC_2H_5^* + C_2H_6$
[DEZ pulse] (Eq. 7)

$ZnOZnC_2H_5^* + H_2O \rightarrow ZnOZnOH^* + C_2H_6$
[H_2O pulse] (Eq. 8)

where asterisks denote the chemical species on the film surface. By the repetition of reactions (Eq. 7) and (Eq. 8), undoped ZnO films can be deposited with atomic layer control on various substrates. The general overall reaction schemes for ZnO:F films (Figure 6b and c) are given as follows:

$ZnOH^* + Zn(C_2H_5)_2 \rightarrow ZnOZnC_2H_5^* + C_2H_6$ [DEZ pulse] (Eq. 9)

Table 2. Sample identification (I.D.), DEZ-H_2O (Zn-O): DEZ-H_2O/HF (Zn-O/F) cycle ratio, growth temperature, film thicknesses after a total of 1200 process cycles as measured by ellipsometry, growth rates, HF pulse portion in one super cycle, and the atomic content of Zn, O, and F in the films as obtained by energy dispersive X-ray analysis (EDX) [65]

Sample I.D.	Cycle Ratio (Zn-O: Zn-O/F)	Growth Temp. (°C)	*t* (nm)	Growth Rate (Å/cycle)	HF Pulse Portion in 1 super cycle (%)	EDX results (at.%)		
						Zn	O	F
ZnO	1:0	140	260±5	2.165	0	50.7±0.1	49.3±0.1	0
ZnO:F{0.2}*	4:1		262±5	2.180	0.10	51.3±0.1	48.7±0.1	Not detected
ZnO:F(0.5)	1:1		281±5	2.343	0.25	51.3±0.1	48.2±0.1	0.5±0.1
ZnO:F(0.7)	1:2		281±5	2.343	0.34	51.2±0.1	48.1±0.1	0.7±0.1
ZnO:F(1.0)	1:4		287±5	2.390	0.40	51.4±0.1	47.6±0.1	1.0±0.1
ZnO:F(1.2)	0:1		301±5	2.510	0.50	51.1±0.1	47.7±0.1	1.2±0.1

* In the case of ZnO:F{0.2}, the F-concentration was estimated by linear plot (Figure 8).

$ZnOZnC_2H_5^* + H_2O + HF \rightarrow ZnOZnOH^* + HF + C_2H_6$
[H_2O/HF pulse] (Eq. 10)

$ZnOZnC_2H_5^* + HF \rightarrow ZnOZnF + C_2H_6$
[F doping in ZnO matrix] (Eq. 11)

When DEZ and H_2O/HF were pulsed, ZnO and ZnO:F layers are formed simultaneously by reactions (Eq. 9), (Eq. 10), and (Eq. 11). Thus, it is clearly the F doping step that supplies the F doping source through the injection sequence of H_2O/HF. Table 2 includes detailed information regarding experimental parameters such as the sample I.D., DEZ-H_2O (Zn-O):DEZ-H_2O/HF (Zn-O/F) cycle ratio, growth temperature, film thickness after a process of 1200 total cycles, growth rates, HF pulse portion in one super cycle, and the atomic content of Zn, O, and F in the films as obtained by energy dispersive X-ray analysis (EDX). The doping concentration of F in the ZnO films could be controlled by the number of F source injections times, as demonstrated in Figure 6.

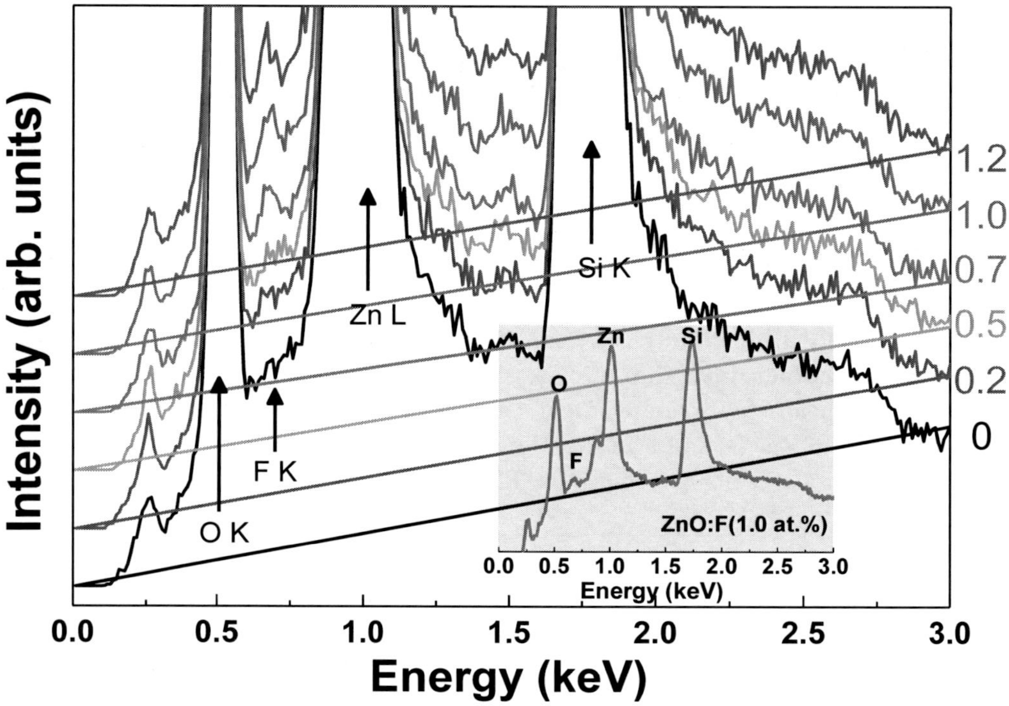

Figure 7. EDX spectra of undoped ZnO and ZnO:F thin films with various F contents. The inset is the EDX results for ZnO:F(1.0) films in logarithmic scale [65].

2-1-1-3. Analyze of the Properties of ZnO and ZnO:F Films

The electrical performance, including the resistivity (ρ) and Hall coefficient (R_H) parameters, were obtained by the van der Pauw method at room temperature using a Hall effect measurement system (Ecopia HMS3000) without the use of a metal soldering for the electrical contacts due to high film quality (DC current = 10 mA) in the magnetic field up to 0.57 T. The film thicknesses were determined by ellipsometry and confirmed by cross-sectional FE-SEM of the cleaved samples. XRD analysis usedCu Kα radiation (λ = 1.5418 Å). For surface morphology analysis, FE-SEM was employed, and quantitative elemental analysis was conducted using EDX with 15 kV energy, 300 sec integration time, and 35° take-off angle (JEOL, JSM 7001F attached with OXFORD X-Max^N EDX detector (Analysis software was AZtecEnergy of OXFORD instryuments.)). The surface elemental analysis of the films was carried out by using X-ray photoelectron spectroscopy (XPS; Thermo VG, U.K.) with monochromated Al X-ray sources (Al kα line: 1486.6 eV) and X-ray power of 12 kV and 3 mA. Room-temperature photoluminescence (PL) spectroscopy (Hitachi P7000) was conducted, and optical transmittance spectra were monitored using an UV-vis-NIR spectrophotometer in the range of 200 to 2500 nm. Grazing-incident wide-angle X-ray diffraction (GIWAXD) used the 9A beam line at Pohang Light Source (PLS, Korea), and photoemission spectroscopy (PES) used the 8A2 beam line at PLS. Au foil was used as the reference material for the calibration of the kinetic energy shift.

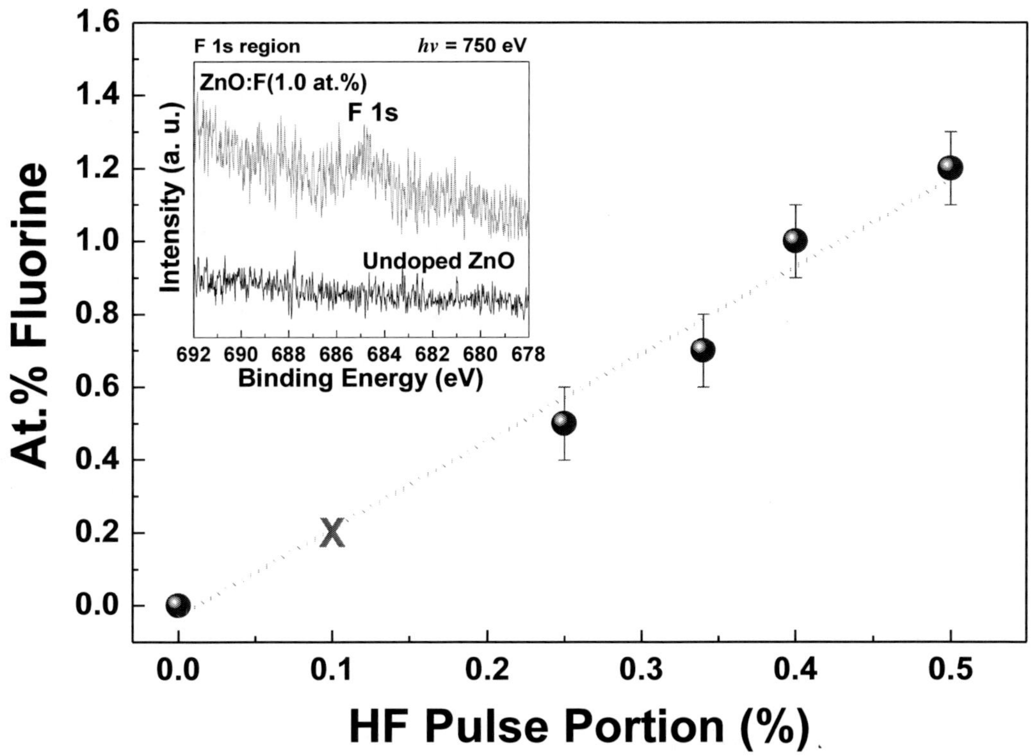

Figure 8. Evolution of atomic fluorine concentration as a function of HF pulse portion. The inset is a high-resolution photoemission spectra of undoped ZnO and ZnO:F(1.0) thin films in the F 1s region [65].

2-1-2. Linearity of F Doping Concentration in ZnO:F Films by ALD

The EDX spectra of undoped ZnO and ZnO:F thin films with various F contents deposited on Si substrates are shown in Figure 7. The intensity of the F signal (peaks at 0.68 keV) was clearly enhanced with an increase in the home-made F source pulse ratio. The inset of Figure 7 is the EDX result of ZnO:F(1.0) films (the representative samples in this work) in logarithmic scale. The plot of the concentration of F in the ZnO matrix as a function of HF pulse portion as measured by EDX is shown in Figure 8. The HF pulse portion in one super cycle could be evaluated by the content of HF in the diluted F source solution (0.5 mL of 48-51% diluted HF in 50 mL water, about 0.5% HF in the source canister) using an HF pulse portion (%) = $[y/(x + y)] \times 0.5\%$. The atomic concentration of F in the films increased linearly with respect to the increase in the HF pulse portion, and the concentration reached 1.2 ± 0.1

at.%. This suggested that the F doping concentration using diethylzinc (DEZ) and home-made F sources was saturated at a value of 1.2 ± 0.1 at.% in the ZnO matrix (See Figure 8). The F concentration of the ZnO:F{0.2} films was undetected by EDX measurements, but the concentration was estimated to be approximately 0.2 at.% based on the linear plot (marked by × in Figure 8). The surface analysis of the films was carried out by using XPS in the F 1s are confirmed from the XPS elemental analysis even though the surface compositions (~0.18 at.% for ZnO:F{0.2} films and ~1.17 at.% for ZnO:F(1.2) films) were very slightly different with the composition for whole depth of the films from EDX. The inset in Figure 9 is the PES F 1s region in the ZnO and ZnO:F(1.0) films. From this, we could also confirm the F doping in the ZnO matrix. A summary of the F concentration in the ZnO films is also shown in Table 2. This linearity between atomic doping concentration of F in ZnO films and HF pulse portion can provide the controllability of F doping concentration by manipulating source pulse ratio of Zn-O and Zn-O/F. In order to monitor the self-limiting conditions at a deposition temperature of 140 ^{o}C and determine the thickness of the films for the optimization of the electrical properties, the thickness and resistivity of the representative ZnO:F thin films (ZnO:F(1.0)) as a function of cycle number is shown in Figure 10. The film thickness increased linearly according to the increase in the number of cycles in the ALD process, and the resistivity decreased significantly up to 1200 total cycles, and increased only slightly thereafter. The lowest resistivity was 1.876×10^{-3} Ωcm in the ZnO:F(1.0) thin films processed with 1200 cycles, and the thickness was 287 ± 5 nm as measured by ellipsometry and FE-SEM. The resistivity values were obtained by Hall effect measurements of the films on glass substrates via the van der Pauw method. The film thickness of undoped ZnO and ZnO:F films with various concentration of F are shown in Figure 11. The growth rates of each film are summarized in Table 2. The growth rates of the films were slightly increased with an increase in the concentration of F due to an enhancement in the crystallinity of the films, and a detailed analysis regarding the increase in growth rate is discussed further with regard to their structural properties.

Table 3. Average crystallite sizes of the samples calculated by using the full-width at half-maximum of (100) and (002) peaks, Hall Effect measurement parameters (carrier concentration (n), mobility (μ), resistivity (ρ), and conductivity (σ)) [65]

Sample I.D.	(100) Crystallite Size (nm)	(002) Crystallite Size (nm)	n (cm^{-3})	μ (cm^2/ Vs)	ρ at RT (Ωcm)	σ at RT (Ωcm)$^{-1}$
ZnO	20.77	21.79	1.713×10^{19}	17.73	2.055×10^{-2}	48.66
ZnO:F{0.2}	19.32	22.32	3.589×10^{19}	43.66	3.983×10^{-3}	251.1
ZnO:F(0.5)	22.90	23.99	5.467×10^{19}	32.49	3.514×10^{-3}	284.6
ZnO:F(0.7)	21.93	22.14	1.243×10^{20}	25.04	2.181×10^{-3}	458.6
ZnO:F(1.0)	22.84	23.71	1.375×10^{20}	24.20	1.876×10^{-3}	533.1
ZnO:F(1.2)	21.81	23.51	1.462×10^{20}	19.49	2.190×10^{-3}	456.5

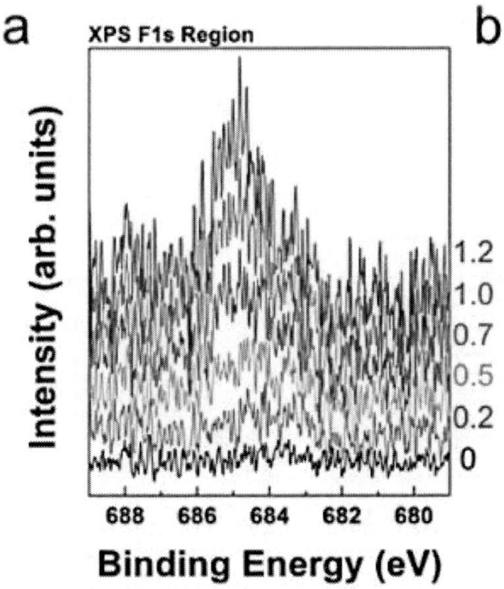

b

ZnO:F{0.2}	Zn $2p_{3/2}$	O 1s	F 1s
BE (eV)	1021.77	530.9	684.5
FWHM (eV)	2.52	2.47	0.58
Area	2784235	559399	2913
A.S.F.	3.354	0.711	1.000
At.%	**51.25**	**48.57**	**0.18**
ZnO:F(1.2)	**Zn $2p_{3/2}$**	**O 1s**	**F 1s**
BE (eV)	1021.77	530.9	684.5
FWHM (eV)	2.47	2.51	0.52
Area	3084872	581590	20534
A.S.F.	3.354	0.711	1.000
At.%	**52.31**	**46.52**	**1.17**

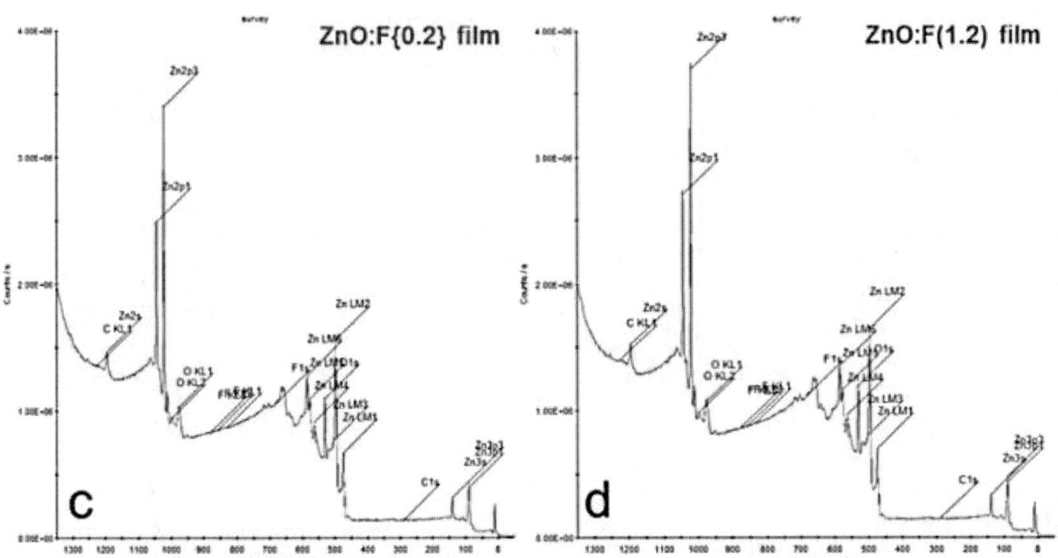

Figure 9 (a). The overlay of XPS spectra for undoped ZnO and various contents of ZnO:F films in F 1s region and (b) the peak-fit results (full-width at half-maximum and atomic sensitivity factor for X-ray sources at 90°) of Zn $2p_{3/2}$, O1s, and F1s peaks for ZnO:F{0.2} and ZnO:F(1.2) samples. The XPS spectra of wide scan for (c) ZnO:F{0.2} and (d) ZnO:F(1.2) films are used for the elemental analysis by using the peak fitting program (Thermo Avantage v3.25) [65].

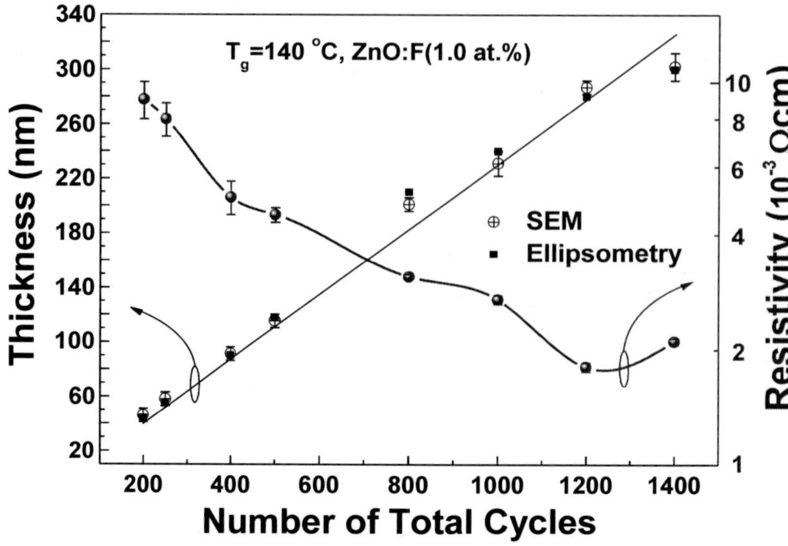

Figure 10. The thickness and resistivity values of ZnO:F(1.0) films deposited on glass substrates at a growth temperature of 140 °C [65].

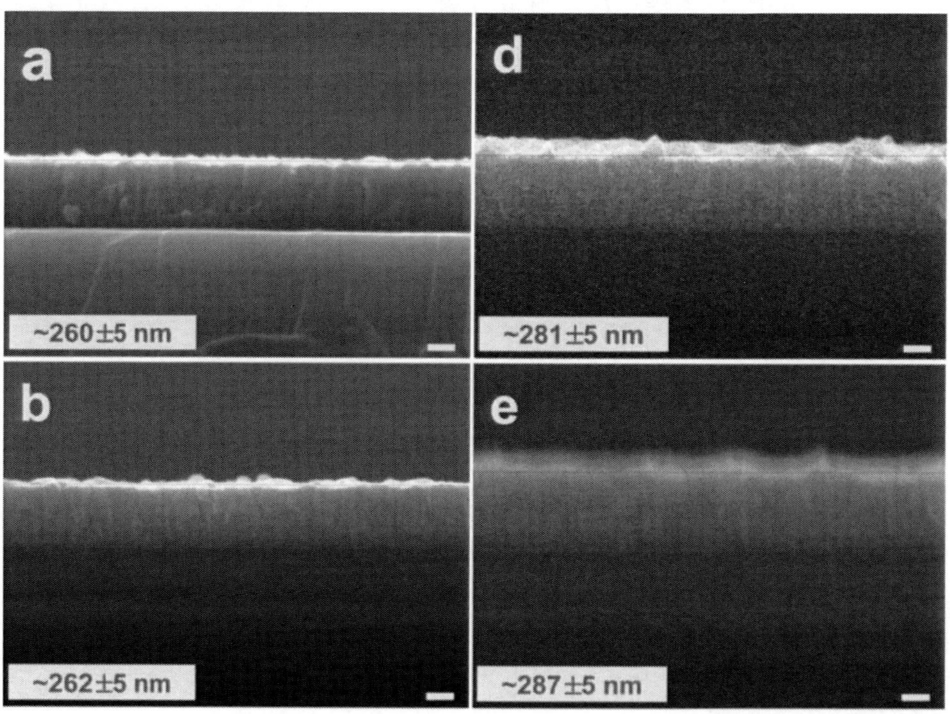

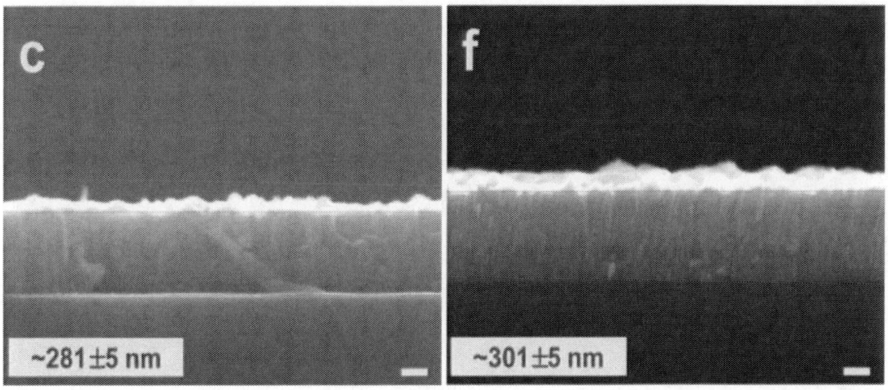

Figure 11. Cross-sectional FE-SEM images of (a) undoped ZnO, (b) ZnO:F{0.2}, (c) ZnO:F(0.5), (d) ZnO:F(0.7), (e) ZnO:F(1.0), and (f) ZnO:F(1.2) films grown at 140 °C on Si substrates (All scale bars represent 100 nm) [65].

2-1-3. Electrical Behaviors of ZnO:F Films by F Doping Concentration

The electrical parameters of ZnO:F films with F concentrations ranging from 0 at.% to 1.2 at.% were determined by Hall effect measurements of the films on glass substrates via the van der Pauw method. Figure 12 a and b show plots of resistivity, Hall mobility, and carrier concentration of the samples as a function of F concentration. The variation in the electrical parameters is summarized in Table 3. With an increase in F concentration, the carrier concentration gradually increased in the range from 1.713×10^{19} cm^{-3} to 1.462×10^{20} cm^{-3}. These concentration values with F doping are well matched with previously reported values in ZnO:F films [68-70] and relatively low when compared with cation doping in ZnO, such as with ZnO:Al and ZnO:Ga, which have higher concentrations of up to $\sim 10^{21}$ cm^{-3} [70, 71]. In the case of cation-doped ZnO films, the electron carriers are simultaneously occupied by Zn substitution of group III ions and native O vacancies, while F ions substitute the O sites in ZnO lattice. Therefore, ZnO:F films have a relatively low concentration compared to cation-doped ZnO films. On the other hand, films doped with cations typically show lower mobilities, under 20 cm^2/Vs, than ZnO:F films [70, 71]. In this work, the mobility of ZnO:F films varied from 19.49 cm^2 V^{-1} s^{-1} to 43.66 cm^2 V^{-1} s^{-1} with F concentration. Due to the low formation energy of the O vacancy, O sites are favorable doping site for F ions, as they have a similar ionic radius to that of O^{2-} [72]. The conduction band of ZnO are composed with the empty 4s levels of Zn^{2+} ions (from ionic characteristics) or with the lowest anti-bonding sp^3 hybrid states (from covalent characteristics). In the same manner, the valence band of ZnO originates from the filled 2p levels of O^{2-} ions or the energetically highest bonding sp^3 states. Therefore, this behavior of substitution for O, rather than Zn, would relatively prevent the perturbation in the conduction band compared with metal dopants, which substitute to Zn. This effect leads to a dramatically increase in mobilities from 17.73 cm^2 V^{-1} s^{-1} for undoped ZnO films to 43.66 cm^2 V^{-1} s^{-1} for ZnO:F{0.2} films at the low doping concentration region. According to an increase of degeneracy of the ZnO:F films with an increase of F doping concentration, however, F dopants could also act as centers of ionized impurity scattering by contributing to an increase in free electrons because the addition of F^- anions raises the Fermi level of ZnO:F films into the conduction band due to filled unoccupied states in conduction bands. Therefore, the mobilities were monotonically decreased from ZnO:F{0.2} films to ZnO:F(1.2) films. Through a combination of carrier concentration and mobility, the lowest resistivity value obtained was 1.876×10^{-3} Ω cm for ZnO:F(1.0) films when the carrier concentration and mobility were 1.375×10^{20} cm^{-3} and 24.20 cm^2 V^{-1} s^{-1}, respectively. The highest resistivity value, as expected, was 2.055×10^{-2} Ω cm for undoped ZnO films and the carrier concentration and mobility were 1.713×10^{19} cm^{-3} and 17.73 cm^2 V^{-1} s^{-1}, respectively. In order to check the effectiveness of F in ZnO matrix as a dopant, the doping efficiency (η_{DE}), was calculated under the assumption that every incorporated F^- ion provides one free electron with substitution to O^{2-} ion. The η_{DE} value is defined as the ratio of the electron concentration to the F atomic concentration in ZnO:F film, which supposed to have the density of 5.606 g cm^{-3} as same as that of the bulk ZnO [68].

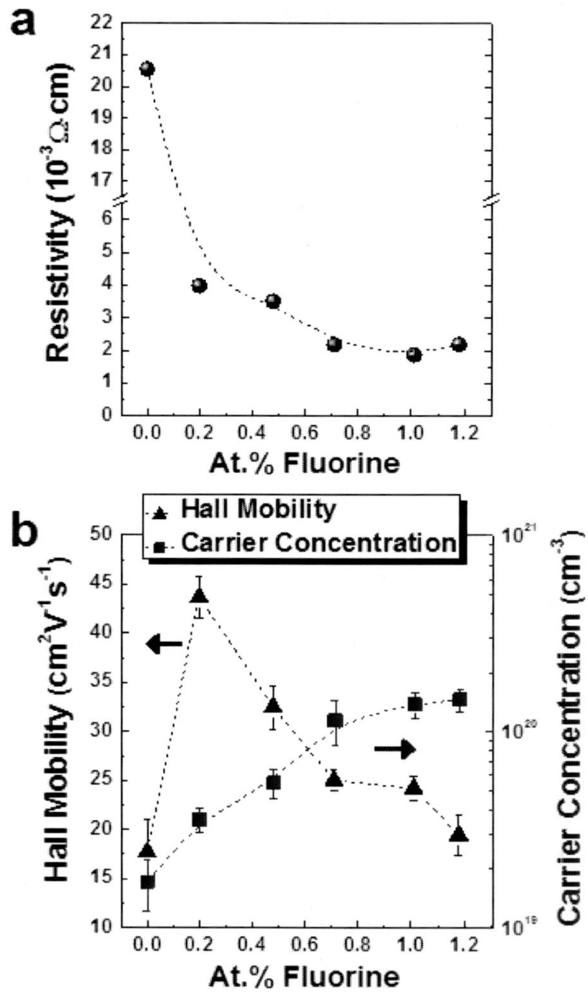

Figure 12. (a) Resistivity of the undoped ZnO and ZnO:F films with various F contents deposited on glass substrates at a growth temperature of 140°C and (b) Hall mobilities and carrier concentrations of the films as a function of atomic fluorine concentration [65].

The physical meaning of η_{DE} provides the fraction of the electrically activated F atoms to F atoms doped in the film. The highest η_{DE} value was 21.6% for ZnO:F{2.0} film and the η_{DE} value of ZnO:F(1.0) film (lowest resistivity in this work) was 16.4%. These values are relatively low than previously reported value as high as 65~97% for CVD-grown ZnO:F films at 375°C ~ 450°C by Hu et al. [68]. This might be resulted by the very high growth temperature than 140°C in this work, which lead to easy In order to check the effectiveness of F in ZnO matrix as a dopant, the doping efficiency (η_{DE}) activation of F dopants in the films. On the other hand, Lee et al. reported the electronconcentration of 1.4 × 10^{20} cm^{-3} and 1.9 × 10^{20} cm^{-3} for ZnO:Al film with 1.9 at.% Al (the lowest resistivity) and with 3.4 at.% Al (the highest carrier concentration) deposited by ALD at 200°C, respectively [73]. From this information, the η_{DE} values of the films could be calculated by 8.7% and 6.5%, respectively, because Al^{3+} ion donates one electron in ZnO:Al film with substitution to Zn^{2+} ion as same as F doping. These η_{DE} value for ZnO:Al films is quite smaller than that of ZnO:F films even if the films were grown at a relatively higher temperature (200°C). These low η_{DE} values of ZnO:Al films are clearly considered by the laminated structure of ZnO and Al_2O_3 alloys from Al doping process by ALD, while a laminated structure is not necessary to incorporate the F atoms into ZnO matrix in this work as mentioned in Introduction section. And also, the effectiveness of F as a dopant material for ZnO films is clearly confirmed from this result.

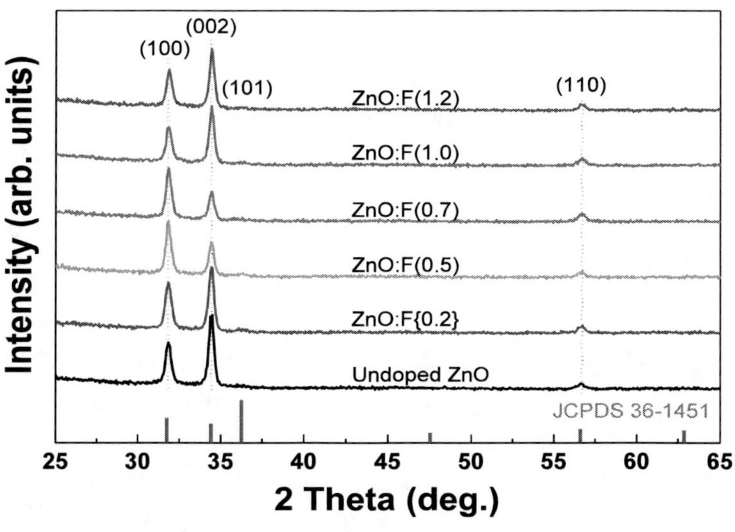

Figure 13. Overlay of XRD patterns of undoped ZnO and ZnO films with various F contents deposited at a growth temperature of 140°C on glass substrates [65].

2-1-4. Growth Behaviors of ZnO:F Films by F Doping Concentration

Crystal structures of undoped ZnO and ZnO:F films with various doping contents grown on glass substrates were analyzed by XRD. The XRD spectra of the films grown by 1200 total ALD cycles at a constant substrate temperature (140°C) are shown in Figure 13. The main diffraction peaks of all films at 2θ angles of 31.74°, 34.46°, and 56.60° were indexed as (100), (002), and (110), respectively, agreed with the standard diffraction pattern (JCPDS 36-1451) of a hexagonal wurtzite ZnO structure. The crystal growth of the ALD-grown ZnO films is mainly influenced by growth temperature [74]. In our experimental conditions, the preferred orientation peaks were a mixture of (100) and (002), and this result was well matched with previous work by Pung et al. [74]. By using the well-known Scherrer equation, the calculated average grain sizes of the samples by using the full-width at half-maximum of

(100) and (002) peaks were summarized in Table 3. The average grain sizes of the sample were ranged from 19 nm to 24 nm without particular tendency according to F doping concentration because the films were deposited at a constant temperature of 140°C. The change in the preferential orientation between (100) and (002), however, may be explained by the Fconentration. At the same deposition temperature, an increase in the F doping concentration from 0 at.% to 0.2 at.% leads to a (002) preferred orientation. With a further increase in the F concentration from 0.5 at.% to 0.7 at.%, the preferred orientation was changed from (002) to (100), and that of the films doped with more than 1.0 at.% was changed from (100) to (002). This change in the preferred orientation of ZnO thin films according to F doping concentration at the same deposition temperature has not been reported previously, as ZnO:F films exhibit a hexagonal structure with a strongly preferred (002) oriented plane due to growth or post-annealing by PLD [75], radio frequency magnetron sputtering [70], electron beam evaporation [69], and CVD [68] at high temperatures of over 450°C. Liu et al. previously reported first-principles studies of ZnO:F and suggested that the most favorable sites for F doping into a ZnO matrix are O vacancies, which have the lowest formation energy and act as the dominant defect species [72]. Therefore, it is anticipated that the growth of the a-axis-oriented (100) plane is more active than the growth of the c-axis-oriented (002) planes perpendicular to the substrate due to F^- anions filling O vacancies or substituting for O sites as the F doping concentration increases. Moreover, ZnO:F films have a saturation point of F doping into the ZnO matrix, and thus it is apparent that the preferred (100) orientation was changed to the preferred (002) orientation as the F doping concentration increased above 0.7 at.%. This fact explained the morphology of the ZnO and ZnO:F films as well as the XRD patterns. A change in peak position resulting from lattice distortion with F doping was not observed because the ionic radius of F^- anion (~1.31 Å) is very similar to O^{2-} anion (~1.38 Å) when F^- anions fill oxygen defect sites.

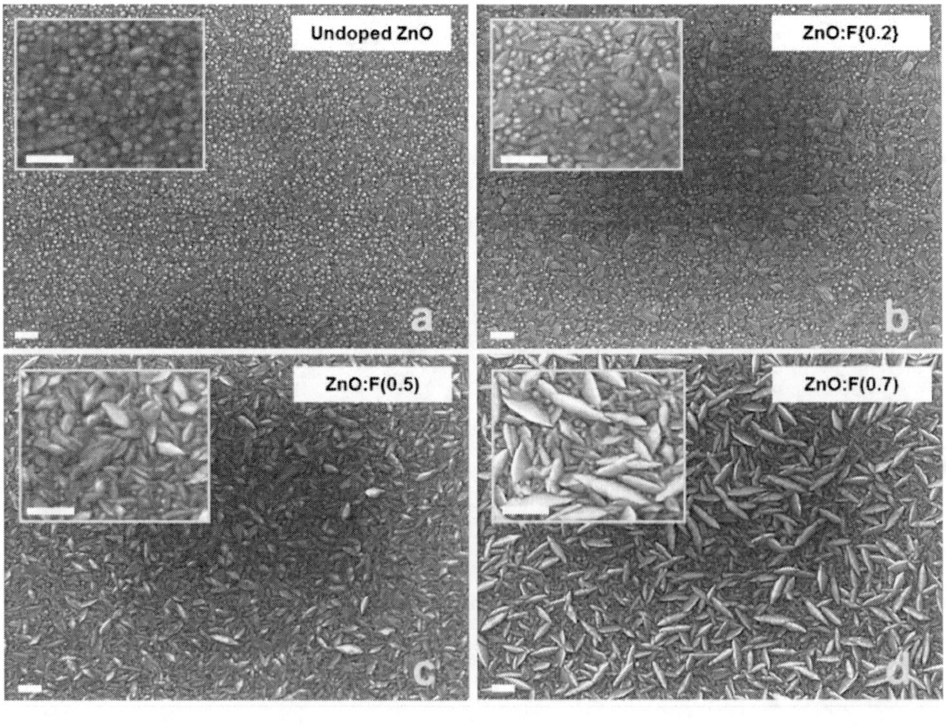

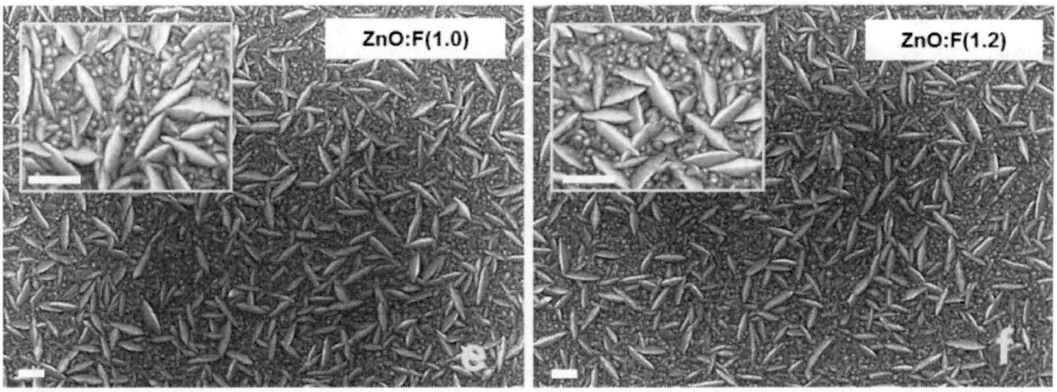

Figure 14. Top view FE-SEM image of (a) undoped ZnO, (b) ZnO:F{0.2}, (c) ZnO:F(0.5), (d) ZnO:F(0.7), (e) ZnO:F(1.0), and (f) ZnO:F(1.2) films grown at 140°C on glass substrates (All scale bars represent 200 nm) [65].

The surface morphology of ZnO and of ZnO:F thin films with various F concentrations deposited on glass substrates by ALD were investigated by FE-SEM, and the resulting images are shown in Figure 14. The undoped and doped ZnO films made at the same deposition temperature (140°C) exhibited a surface structure consisting of grains of different size and morphology that were strongly dependent on the F doping concentration. Normally, columnar and wedge-like morphologies are regarded as c-axis and a-axis growth orientations, respectively [76]. Undoped ZnO thin films grew with a c-axis orientation. With an increase in F content, the c-axis-oriented grains decreased and the a-axis-oriented grains increased up to ZnO:F(0.7). When the F-doping concentration increased to 0.7 at.%, the grains of the films were largely wedge-like in shape, suggesting that the (100) planes are further growth than (002) planes, which is consistent with the XRD analysis. Moreover, it seems that these phenomena were saturated in ZnO:F(0.7) thin films. Columnar grains increased from ZnO:F(1.0) to ZnO:F(1.2). This morphological change with respect to F doping concentration might be a result of the F doping mechanism, and is not influenced by difference in grain growth with temperature as films were grown at a constant temperature. There was good evidence that F^- anions could passivate surface dangling bonds. In fact, in nanocrystalline thin films of ZnO, the high number of surface dangling bonds can act as trap sites for free electron carriers and therefore decrease the mobility, which subsequently leads to an increase in resistivity. Therefore, this passivation effect, through saturating these surface dangling bonds by F^- anions at grain boundaries, might result in a drastic increase in the mobilities from 17.73 cm^2/Vs for undoped ZnO films to 43.66 cm^2/Vs for ZnO:F{0.2} films with a slight increase in carrier concentration from intrinsic ZnO to ZnO:F{0.2} films. In solar cell applications, a textured transparent electrode surface is a necessary precondition for light trapping, as this enhances the light absorption and solar efficiency of active cells. In current commercial applications, the surface texture is developed by the post-deposition wet-etching of smooth, deposited ZnO films [77]. The approach of fabricating textured ZnO:F films presented here has potential for use in light-trapping applications through a single-step process without wet-chemical etching, and may also be adopted for flexible substrates that require a low deposition temperature. There are limitations of the analysis based on the XRD patterns and surface morphologies of the films in understanding the growth modes of ZnO:F films according to increasing F doping concentration. Thus, a GIWAXD study was performed to better understand the growth mechanism because the information depth in a GIWAXD study could totally cover the film thickness of ZnO:F films according to the growth direction (Si substrate peaks could be detected in the experiments). Figure 15 shows the GIWAXD patterns of the 2D image and 1D scattering intensity curves (the detection angles are 10°, 45°, and 90° from the Si substrate) for (a) undoped ZnO and (b-f) various ZnO:F films grown at 140°C on Si substrates. To determine the 2D intensity distribution, the horizontal (or out-of-plane) component of the scattering vector q_y was plotted along the *x*-axis, and the vertical (or in-plane) component of the scattering vector q_z was plotted along the *y*-axis. The observed diffraction peaks were addressed according to the Miller indices (*hkl*). When the growth component perpendicular to the substrate was considered, the growth mode of the undoped ZnO films (Figure 15a) showed a similar behavior to that suggested by the XRD results. However, with an increase in the F doping concentration, the growth of (100) planes increased up to an F concentration of 0.7 at.% (Figure 15c), after which the growth of the (100) planes was saturated. When the concentration of F exceeded 0.7 at.%, the intensity of the (100) planes did not change, then the growth of (002) planes increased up to the maximum F doping concentration (ZnO:F(1.2) films). Based on these results, the assumed growth mode of the ZnO:F films as determined via XRD and morphological analysis could be confirmed. F^- anions preferentially substitute for O^{2-} anion sites and O vacancies with the growth of (100) planes, and preferential filling continues with the growth of (002) planes in the longitudinal direction of the hexagonal wurtzite structure as the F doping concentration increases. Therefore, it is suggested that diminishing O-related defect sites in ZnO:F films leads to the enhancement of the growth rate of the films.

2-1-5. Optical Behaviors of ZnO:F Films by F Doping Concentration

In order to confirm the structural-defect-related properties associated with the aforementioned growth mode, room temperature PL spectra for undoped ZnO and representative ZnO:F(1.0) thin films grown at 140°C on Si substrates, which showed the minimum resistivity, were monitored and are shown in Figure 16. Both films exhibited a strong luminescence peak at ~379 nm for undoped ZnO and ~377 nm for ZnO:F(1.0) films, the position of which is known as the near-band-edge (NBE) transition. A broader emission band centered at ~500 nm was located in the green emission portion of the visible region, which can be explained by deep level defects such as O vacancies and Zn interstitials in the ZnO crystals [78, 79]. Therefore, the significant abatement of the intensity of visible emissions is likely a result of the decrease in O vacancies, which can be attributed to the F^- anions that occupy the empty sites in the ZnO matrix [80]. This result is well matched with previous PL analyses of ZnO:F films [80, 81] and the results the XRD and GIWAXD analyses presented in this work. In order to confirm the enhancement of crystal quality due to F doping, the I_{UV}/I_{VIS} ratio was calculated, where I_{UV} and I_{VIS} are the intensities of the UV and visible emission peaks, respectively. The calculated values of I_{UV}/I_{VIS} for undoped ZnO and ZnO:F(1.0) films were 1.740 and 2.609, respectively. This indicated that the ZnO:F(1.0) films had better crystallinity and fewer defects than the undoped ZnO films due to the substitutional doping of O ions with F ions and the filling of O-related defect sites. This resulted in increase in growth rate with increases in F concentration in ZnO crystals, even at low temperatures.

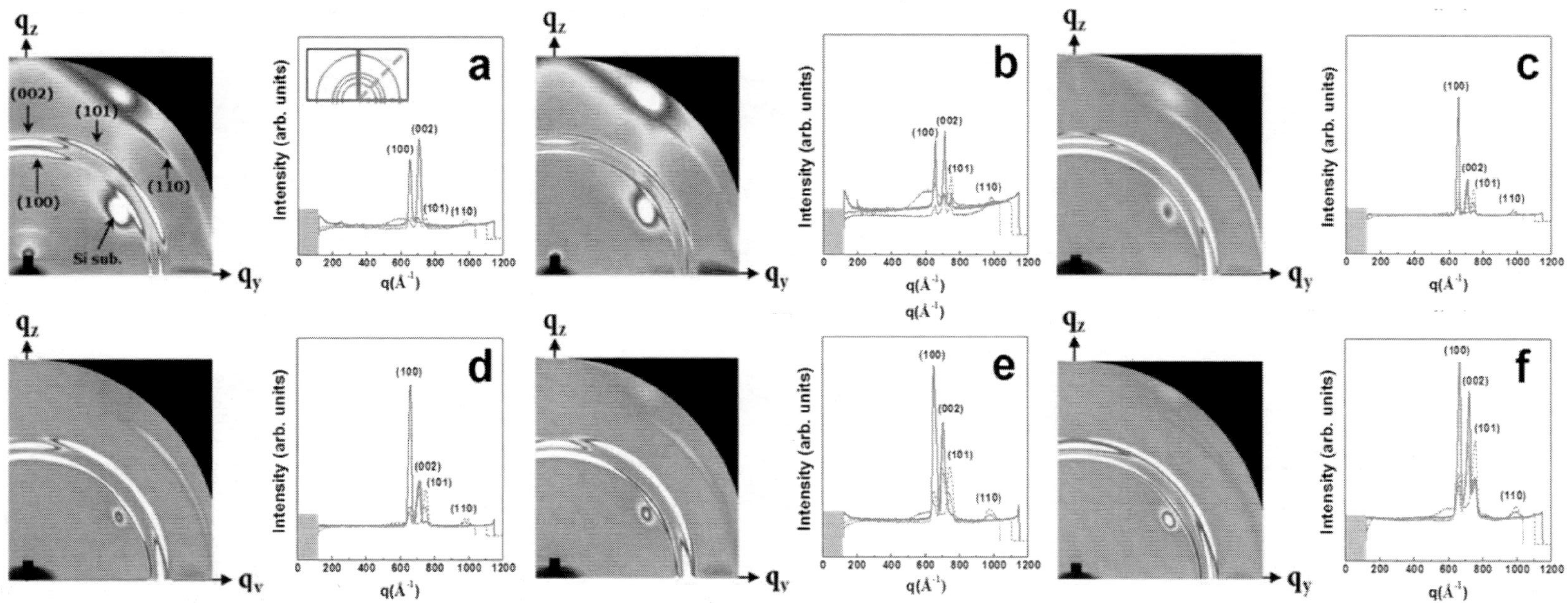

Figure 15. GIWAXD patterns of the 2D image and 1D scattering intensity curves of (a) undoped ZnO, (b) ZnO:F{0.2}, (c) ZnO:F(0.5), (d) ZnO:F(0.7), (e) ZnO:F(1.0), and (f) ZnO:F(1.2) films grown at 140°C on Si substrates. For the 2D intensity distribution, the horizontal (or out-of-plane) component of the scattering vector q_y is plotted along the *x*-axis, and the vertical (or in-plane) component of the scattering vector q_z is plotted along the *y*-axis. The observed diffraction peaks are addressed according to Miller indices (*hkl*) [65].

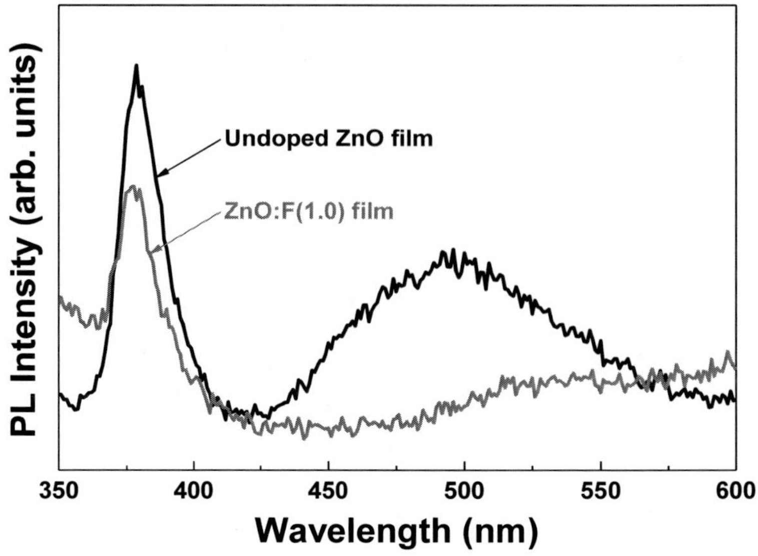

Figure 16. PL spectra of undoped ZnO and ZnO:F films deposited on glass substrates at 140°C. The spectra were measured at room temperature with a laser excitation wavelength of 325 nm [65].

Figure 17 shows the optical transmittance of undoped ZnO and various ZnO:F films from 200 nm to 2500 nm (Figure 17a) and the visible range from 200 nm to 1000 nm (Figure 17b). In TCOs, the three main regions of transmittance are typically divided into UV, visible, and NIR. As the carrier concentration of the films increased through F doping in ZnO, the absorption was increased in the NIR region. This phenomenon can be explained by the classical Drude free-electron theory. In other words, absorption by free electrons restricts the transmittance of films with high electron density in the conduction band such that the absorption and reflection of electron gas (plasmon) increases and the transmittance decreases for frequencies lower than the plasma frequency (for wavelengths higher than the plasma wavelength) as follows:

$$\omega_P{}^2 = \frac{ne^2}{\varepsilon_{ZnO}\varepsilon_0 m_e^*} \quad \text{(Eq. 12)}$$

$$\lambda_P = \frac{c}{\omega_P} \quad \text{(Eq. 13)}$$

where ω_P is the plasma frequency; *n* is the electron concentration; *e* is the electron charge; ε_{ZnO} and ε_0 are the dielectric constant measured in the transparent region of the spectrum of undoped ZnO and the vacuum permittivity; λ_P is the plasma wavelength; and c is the velocity of light in a vacuum [82]. Therefore, with an increase in the carrier concentration with an increase in the F doping concentration, the absorption of the films was increased by the shortening of the plasma wavelength. These results are well matched with the electrical properties of ZnO and the various ZnO:F films. In the literature [83], the plasma wavelengths of ZnO:F and ZnO:Al (representative cation-doped ZnO) films were reported to be greater than 2000 nm and 1300 nm, respectively.

Therefore, this result demonstrated the superiority of ZnO:F films compared with cation-doped ZnO films; namely, the higher mobility of the ZnO:F films led to low resistivity even though the carrier concentration was lower than in cation-doped ZnO films. This yields films that are suitable for applications in solar cells for long wavelength light absorption. The average transmittance values in the visible light region (from 380 to 780 nm) of the undoped ZnO, ZnO:F{0.2}, ZnO:F(0.5), ZnO:F(0.7), ZnO:F(1.0), and ZnO:F(1.2) films were 79.93%, 81.01%, 80.58%, 81.76%, 82.99%, and 83.03%, respectively. Similarly to other reported results [70], the enlarged transmittance of the visible range for undoped ZnO and ZnO:F films with various F optical transmittance of the ZnO:F films in the range from 380-780 nm was slightly enhanced with increased F content. This result could be explained by a more facile electron transition for absorbing photon energy in the short wavelength region with increasing F anion concentration in the ZnO matrix. Furthermore, the occupation of O^{2-} sites by F^- anions could result in a decline in O vacancies, which could decrease absorption and emission in the visible region as a result of the O defects. The transmittance of the films on glass substrates was consistently higher than 80%, which is suitable for transparent electrodes.

The visible region from 500 nm to 550 nm is known to be the region of maximum absorption in organic solar cells, and the transmittance of all ZnO:F films reached nearly 90%, which is the same as an LCD glass substrate. Therefore, the ZnO:F thin film developed in this work could possibly serve as transparent electrodes for organic solar cells due to the high transparency at the maximum absorption wavelength.

To further investigate the optical behaviors of ZnO:F thin films in the UV region, the absorption coefficient was calculated using Lambert's formula:

$$\alpha = -\frac{1}{t}\ln \mathrm{T_r} \qquad \text{(Eq. 14)}$$

where t and T_r are the film thickness and transmittance, respectively [84]. The direct band gaps of the films were estimated from the $(\alpha hv)^2$ vs. hv (Tauc relation) plot by extrapolating the fit of the linear region to $\alpha = 0$, shown in the inset of Figure 17b. The estimated optical band gap values increased with increasing F content and the values of the undoped ZnO, ZnO:F{0.2}, ZnO:F(0.5), ZnO:F(0.7), ZnO:F(1.0), and ZnO:F(1.2) films were 3.277 eV, 3.299 eV, 3.323 eV, 3.351 eV, 3.400 eV, and 3.414 eV, respectively. Furthermore, the shift in the absorption edge increased with increasing amounts of F. The band gap of the ZnO film, 3.277 eV, was in agreement with the ideal band gap of pure ZnO. The increase in the band gap with respect to increasing F content is generally believed to be due to the Burstein- Moss effect [85, 86]. ZnO films are n-type oxide semiconductors due to the electron donors generated by O vacancies and Zn interstitials. The addition of donor F^- anions raises the Fermi level of the ZnO:F films into the conduction band, causing the films to completely degenerate, and the absorption edge therefore moves to an energy higher than the actual band gap of pure ZnO. Finally, the monotonic increase in the absorption in the NIR region and the optical band gap in the UV region could be considered to be additional evidence of the increase in electron concentration as the F doping concentration increases. In order to verify the optical band gap widening in ZnO:F films by BM effect, the dependence of the bandgap shift on electron concentration, $n^{2/3}$ is presented in Figure 18. Assuming parabolic energy band, and due to direct-band gap of ZnO, the optical band gap widening can be presented by the energy [87]

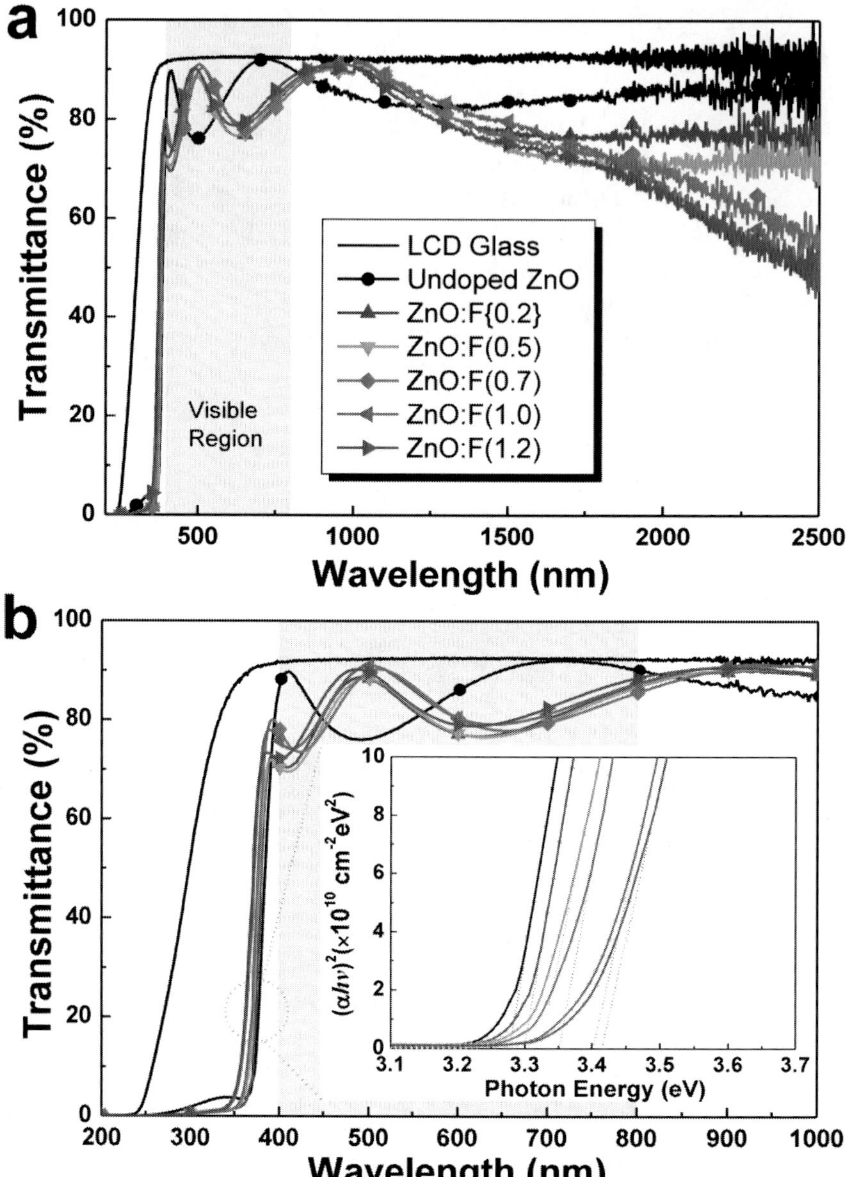

Figure 17. (a) The optical transmittance ranging from 200 nm to 2500 nm of undoped ZnO and ZnO:F films with various F contents deposited on glass substrates at 140°C. (b) The contents. The inset is a plot for the evaluation of the optical band gap of the films [65].

$$\Delta E_{BM} = \left(\frac{h^2}{8m^*}\right)\left(\frac{3}{\pi}\right)^{2/3} n^{2/3} \qquad \text{(Eq. 15)}$$

where h is Plank's constant, m^* is the reduced effective mass, and n is the electron concentration in ZnO and ZnO:F films. In Figure 18, the optical band gap shift is observed to be directly proportional to $n^{2/3}$ by linear fitting of the experimental results. The slope of the linear fitting line yields $m^* = 0.62m_0$, where m_0 is the free electron mass in free space. This m^* value is similar to the previously reported one, which are $0.60m_0$ for ZnO:F films [75] and $0.55m_0$ for ZnO:Ga films [88]. This m^* value is higher than the thoeritical reduced effective mass of $0.19m_0$ ($1/m^* = 1/m_e^* + 1/m_h^*$, where m_e^* and m_h^* present the electron effective mass ($0.28m_0$) and hole effective mass ($0.59m_0$)) for ZnO [89]. The enhancement of the effective mass might be caused by a possible nonparabolicity effect of the ZnO conduction band and also, band gap renormalization with the degeneracy of the films by an increase of electron concentration in conduction band [90].

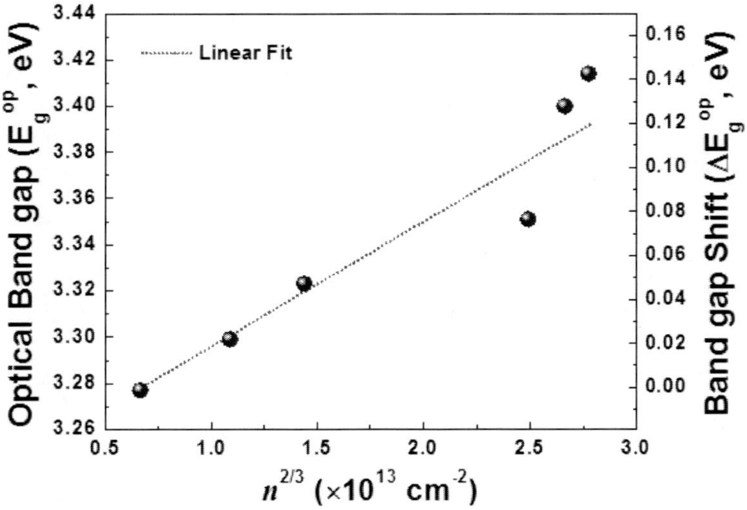

Figure 18. The dependence of the optical band gap and band gap shift of the films on the carrier concentration ($n^{2/3}$). The dotted line represnets the linear fit of experimental results [65].

2-2. Band Gap Widening and Band Gap Narrowing Effects in ZnO:F Thin Films

Zinc oxide (ZnO), which is an n-type, wide band gap (3.3 eV) semiconductor with a high exciton binding energy (60 meV), is the most promising TCOs due to its potential optoelectronic applications, such as solar cells [91, 92], thin film transistors [93], light-emitting diodes [94], and flat panel display [95]. Therefore, it is very important to understand the optoelectronic properties of ZnO for application to those electrical devices as a transparent electrode. Many groups have reported the discussion about the difference between the band gap shift from pure band filling model and actual band gap shift of impurities-doped ZnO thin films but it is relatively weak compared with studies on the performances of TCOs, such as resistivity, mobility, and conductivity. And also, most of researches have been focused on ZnO with group III dopants (Al [96], Ga [97], B [98], and In [99]). But the most effective dopant is known by F because the ionic radius of F^- (~1.36 Å) is close to that of O^{2-} (~1.40 Å). Thus, compared with group III elements, a lower lattice distortion could be expected for F doping into ZnO matrix. Moreover, F-doped ZnO (ZnO:F) thin films could significantly improve the mobility of carriers and therefore F is appropriate anion dopant candidate for application of transparent electrode in optoelectronic devices. With the carrier concentration increase by doping in ZnO above critical carrier concentration of N_C, ~2-3 ×

10^{19} cm^{-3} in ZnO films [100], ZnO films show the degenerate semiconductor behaviors, which are band gap widening and narrowing effects. Band gap widening effect results in the increase of optical band gap (E_g^{op}) of degenerately doped ZnO more than nondegenerate ZnO films. This phenomenon is caused by the well-known BM effect (blue shift in absorption spectra) in heavily doped ZnO films. And also, the band gap narrowing occurs due to the band gap renormalization (BGR) effect by many-body interactions including perturbation and carrier-ion correlations in conduction band. Consequently, the measured optical band gap can be compensated by band gap widening and narrowing as $E_g = E_{g0} + \Delta E_{BM} - \Delta E_{BGR}$, where E_{g0}, ΔE_{BM}, and ΔE_{BGR} are the band gap of the undoped ZnO, the energy shift by BM and BGR effects, respectively. Thus, it is important to better understand the relative contributions of band gap widening (by BM effect) and the band gap narrowing (by BGR effect) to control the optical behaviors of doped ZnO and band alignment of the ZnO-based thin films with active materials in the optoelectronic devices. On the other hand, ZnO:F thin films can be deposited by many techniques such as spray pyrolysis [101], radio frequency magnetron sputtering [102], electron beam evaporation [103], and puled laser deposition [104]. Although ALD is a thin film growth method allowing low temperature growth, better step coverage, good uniformity and controllability of thickness, it has not been yet reported the formation of ZnO:F thin films by ALD.

ZnO:F thin films were fabricated by ALD through an use of novel home-made F source. We determined the band gap widening with BM effect and the band gap shrinkage with the interaction between donors and host crystal and compared measured values with proper theoretical models by examining the absorption band shift of ZnO films with an increase of carrier concentration from 1.524×10^{19} cm^{-3} to 1.419×10^{20} cm^{-3} by F doping.

2-2-1. Experimental Details

Around 70-nm-thick ZnO and ZnO:F thin films were prepared on glass (Fusion 1737) substrates via ALD using a traveling wave type Lucida D100 system (NCD Technology, Inc. Korea) at a deposition temperature of 150°C and a working pressure of 3 to 3.3 Torr. Diethylzinc (DEZ, Hansol Chemical Co., Ltd., Korea) and DI water were used as a Zn precursor and a reactant source for O, respectively. F doping into ZnO was performed by using a mixture of DI water 50 mL and diluted hydrogen fluoride (HF 48~51% diluted in water) 0.5 mL as a single reactant source for O and the F dopant. For the control of F doping concentration in ZnO thin films, the cycle ratio of DEZ/H_2O to DEZ/H_2O:HF was controlled 4:1, 1:1, and 1:2. The details of the film preparation with manipulating F doping concentration was previously described. The concentration of F in ZnO in terms of atomic percent could be determined with 0.1 at.%, 0.5 at.%, and 0.7 at.% for 4:1, 1:1, and 1:2 samples, respectively, from quantitative analysis by using EDX with 15 kV energy, 300 s integration time, and 35° take-off angle (JEOL, JSM 7001F attached with OXFORD X-Max^N EDX detector) and XPS with monochromated Al X-ray sources (Al kα line: 1486.6 eV). For the deposition of ZnO and ZnO:F thin films, total ALD cycle was 300 cycles for each samples and the deposition rate was approximately 2.3 Å for cycle. The film thicknesses are E = 8.04621 KeV. For considering surface morphology, FE-SEM was used. PES and valence band spectra were measured and analyzed by 8A2 beam line determined by an ellipsometer and FE-SEM of cleaved samples. The cycle ratio, F concentration (C_F), film thicknesses (t), substrate temperature (T_s), carrier concentration (n), and crystallite size of ZnO and ZnO:F thin films are summarized in Table 4. XRD measurement were carried out using 10B beam line with at Pohang Light Source (PLS) with λ = 1.5409 Å, at Pohang Light Source (PLS). Au foil was used as a reference material for the calibration of the kinetic energy shift. Room-temperature PL was monitored by photoluminescence spectroscopy (Hitachi P7000) and the E_g^{op} was determined from the $(\alpha h\nu)^2$ vs. $h\nu$ (photon energy) using extrapolation of the obtained absorption coefficient the α by an UV-vis-NIR spectrophotometer. The carrier concentration of the films were obtained by the van der Pauw method without metal as electrical contacts due to highly quality films at room temperature using a Hall effect measurement system (Ecopia HMS3000) with the direct current (I_{DC} = 10 mA) four probe method in the magnetic field up to 0.57 T.

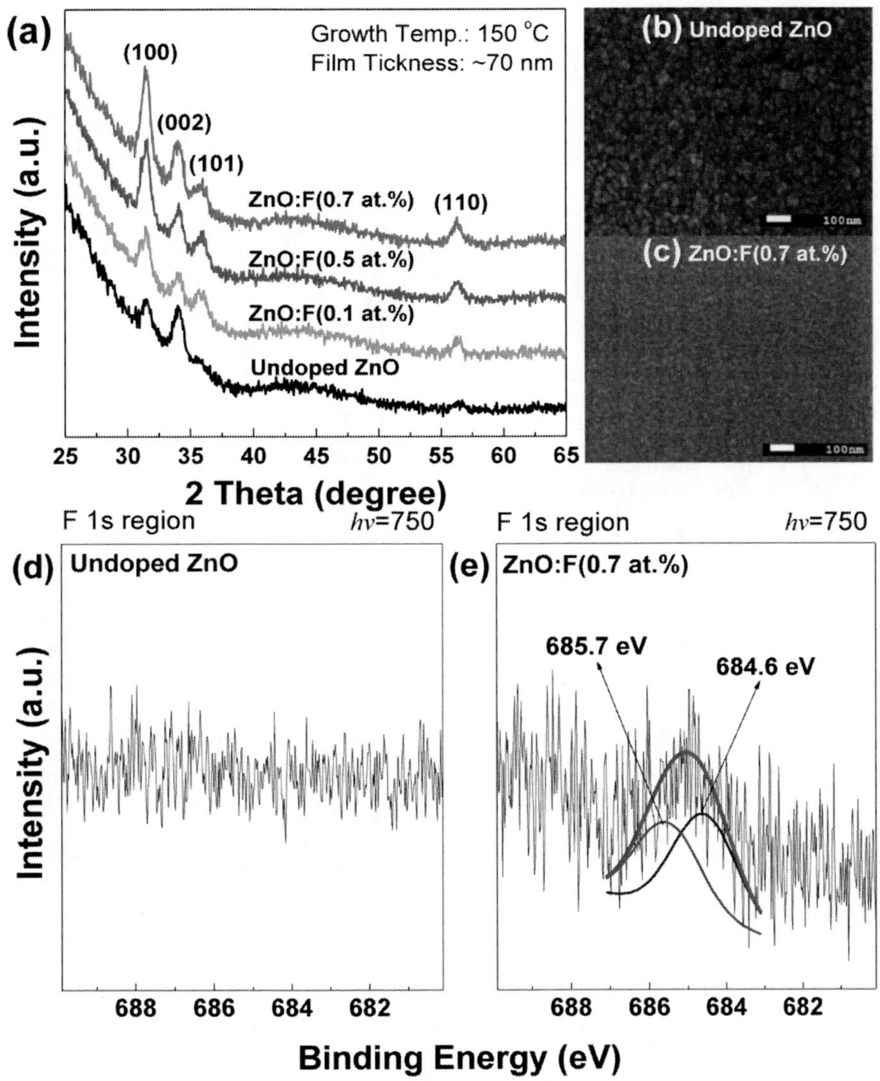

Figure 19. (a) XRD patterns of undoped ZnO and various contents of ZnO:F thin films and the top-view images of FE-SEM for (b) undoped ZnO and (c) ZnO:F(0.7 at.%) films. PES spectra in the F 1s regions of the nanocrystalline (d) undoped ZnO and (e) ZnO:F (0.7 at.%) films [105].

2-2-2. Extrinsic F Doping Effect on Electronic Band Structure of ZnO Thin Films

The XRD patterns of undoped ZnO and ZnO:F thin films on glass substrates and the top-view FE-SEM images of undoped ZnO and ZnO:F(0.7 at.%) films are shown in Figure 19a, b, and c. All films exhibited hexagonal wurtzite structures with visible (100), (002), (101), and (110) peaks. Predominant peak was changed from (002) to (100) with an increase of F doping concentration and the peak position was not changed with F doping into ZnO matrix. It is reveal that ionic radius of F anion (F^-: ~1.36 Å) is much similar to that of O anion (O^-: ~1.4 Å), therefore a lower lattice distortion occurred in ZnO:F thin films compared with cation-doped ZnO films [103]. And also, an increase of F doping concentration from 0.1 at.% to 0.7 at.%, the crystallinity of the films are slightly increased due to fill the oxygen vacancies with F anions. According to the first-principles calculations by Liu and Janotti et al. [106, 107], oxygen vacancy has the lowest formation energy and acts as the dominant defect species under Zn-rich and O-poor condition. Therefore, less energy is needed for F anions entering into the site where the oxygen vacancy forms. The detailed analysis about growth modes of ZnO:F films was previously described. Figure 19d and e show PES spectra in the F 1s regions of undoped ZnO and ZnO:F(0.7 at.%) thin films, respectively. As expected, F 1s peak is not visible in the undoped ZnO films but it appears in ZnO:F(0.7 at.%) films. The line shape of the F 1s peak is slightly asymmetric and the Gaussian line shape fitting suggests two possible bands that are overlapped with each other at the position of 684.6 eV and 685.7 eV, respectively. Two decomposed peaks indicate that F anions may have locally distinct chemical states. The peak at 684.6 eV is attributed to the substitutional site in ZnO matrix, while the peak at 685.7 eV is related to the F anions associated with dangling bonds on the surface of ZnO nanocrystal or grain boundaries [101]. In the Figure 20a, the blue shift of E_g^{op} could be confirmed clearly with an increase of carrier concentration. The absorption coefficient of all films on glass substrate was calculated using Lambert' formula,

$$\alpha = -\frac{1}{t}\ln T_r \qquad \text{(Eq. 16)}$$

where T_r and t are the transmittance and film thickness, respectively [108]. The direct band gaps of the films were estimated from the $(\alpha h\nu)^2$ *vs.* $h\nu$ (Tauc relation) plot by extrapolating a fit of the linear region to $\alpha = 0$ in Figure 20a. The E_g^{op} value for undoped ZnO thin films was 3.27 eV and the values of ZnO:F (0.1 at.%), ZnO:F (0.5 at.%), and ZnO:F (0.7 at.%) were 3.30, 3.38, and 3.41 eV, respectively. The E_g^{op} was widened from 3.27 eV (undoped ZnO) to 3.41 eV (ZnO:F (0.7 at.%)) with an increase carrier concentration of the films from1.524 × 10^{19} cm^{-3} to 1.419 × 10^{19} cm^{-3}. The overall E_g^{op} widening is due to the BM effect. The BM effect comes into play in a n-type degenerated semiconductor, when the electron concentration (n) goes above the critical value of $N_C = 2(2\pi m^* k_B T)^{3/2}/h^3$, where h, k_B, and m^* is the Plank's constant, Boltzmann constant, and the electron effective mass in ZnO, respectively. In case of our ZnO thin film, $N_C = 1.524 \times 10^{19}$ cm^{-3}, and for all ZnO:F films, n is more than N_C as shown in Table 4. This carrier concentration of nondegenerate ZnO films similarly reported by Roth et al. (~2-3 × 10^{19} cm^{-3}) [100]. Due to the BM effect, as Fermi level moves up into the conduction band, the lower states in the conduction band filled up leading to a widening of the E_g^{op},

Table 4. The sample I.D., cycle ratio for fabrication of ZnO:F films, substrate temperature (Ts), doping concentration, film thickness, electron concentration, and crystallite size along with (100) and (002) planes were summarized [105]

Sample I.D.	Cycle Ratio (Zn-O: Zn-O/F)	Ts (°C)	C_F (at. %)	*t* (nm)	*n* (cm^{-3})	(100) Crystallite Size (nm)	(002) Crystallite Size (nm)
Undoped ZnO	-	150	0	67	1.524×10^{19}	18.72	19.49
ZnO:F (0.1 at.%)	4:1		0.1	69	5.83×10^{19}	19.32	21.55
ZnO:F (0.5 at.%)	1:1		0.5	70	9.87×10^{19}	20.21	22.11
ZnO:F (0.7 at.%)	1:2		0.7	73	1.419×10^{20}	21.65	21.34

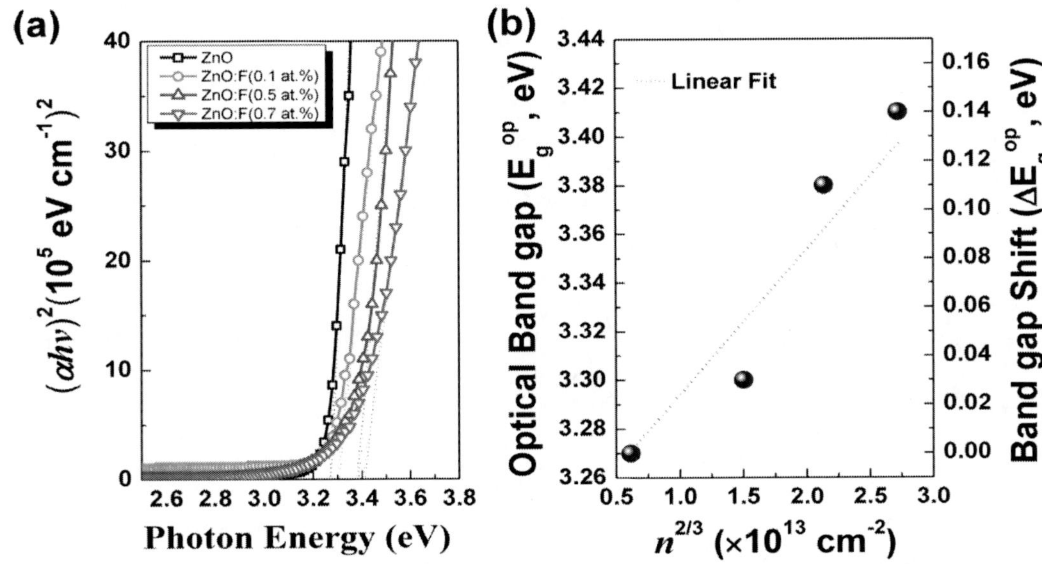

Figure 20. (a) Plot of $(\alpha h\nu)^2$ vs. $h\nu$ for the undoped ZnO and various contents of F-doped ZnO thin films. The linear fit for the optical band gap energy is shown as a dotted line. (b) The dependence of the optical band gap and band gap shift of the films on the carrier concentration ($n^{2/3}$). The dotted line represnets the linear fit of experimental results [105].

$$\Delta E_{BM} = E_F - E_{CBM} = \left(\frac{\hbar^2}{2m_{eh}^*}\right)(3\pi^2 n)^{2/3} \quad \text{(Eq. 17)}$$

where m_{eh}^*, E_F, E_{CBM}, and n are the electron-hole reduced mass, Fermi level, conduction-band minimum, and carrier concentration, respectively. According to the pure band filling model, the E_g^{op} shift should be proportional to electron concentration, $n^{2/3}$. In order to verify the E_g^{op} widening in ZnO:F films by BM effect, the dependence of the band gap shift on electron concentration, $n^{2/3}$ is presented in Figure 20b. Assuming parabolic energy band, and due to direct-band gap of ZnO, the E_g^{op} widening can be presented by the energy [109]. In Figure 20b, the E_g^{op} shift is observed to be directly proportional to $n^{2/3}$ by linear fitting of the experimental results. The slope of the linear fitting line yields $m^* = 0.58m_0$, where m_0 is the free electron mass in free space. This m^* value is similar to the previously reported one, which are $0.60m_0$ for ZnO:F films and $0.55m_0$ for ZnO:Ga films [110]. This m^* value is higher than the theoretical reduced effective mass of $0.19m_0$ ($1/m^* = 1/m_e^* + 1/m_h^*$, where m_e^* and m_h^* present the electron effecive mass ($0.28m_0$) and hole effective mass ($0.59m_0$) for ZnO [111]. The enhancement of the effective mass might be caused by a possible nonparabolicity of the ZnO conduction band and also, BGR with the degeneracy of the films by an increase of electron concentration in conduction band [112].

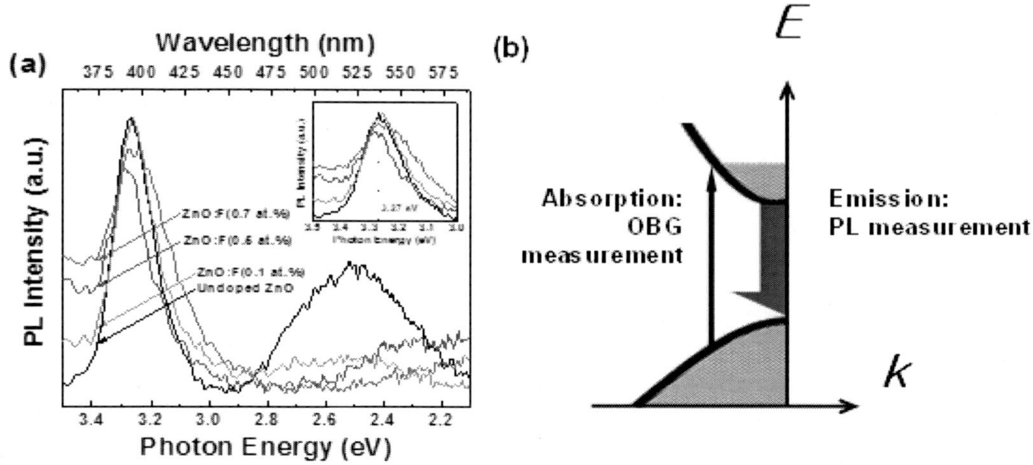

Figure 21. (a) Room-temperature PL spectra of undoped and various contents of ZnO:F thin films (The inset shows the NBE transition region for undoped ZnO and various contents of ZnO:F thin films). (b) Illustration of the optical process in degenerately doped ZnO films and corresponding techniques of measurement in this work. Absorption and emission energies are measured by optical band gap (OBG) shift and PL analysis, respectively [105].

For the calculation of the energy shift via BGR (band gap shrinkage), room-temperature PL measurement was accomplished with the samples of undoped and various contents of ZnO:F thin films are shown in Figure 21a. The inset of Figure 21a shows the enlarged region of near-band-edge (NBE) transition for the thin films. The band gap narrowing from BGR can be caused by the many-body effects including with electron exchange interaction, minority carrier correlations, and carrier-ion correlation, but ΔE_{BGR} can be calculated by the assumption of the value of NBE transition. Because the NBE transition of direct bang gap semiconductors in PL measurement should be occurred by emission process of band-to-band relaxation as shown in Figure 21b. However, in our experimental condition, band gap shrinkage by Stokes shift was not observed with an increase of carrier concentration by F doping. Generally, red shift presented by band gap narrowing from an increase carrier concentration of conduction band of ZnO films in NBE transition region. There was, however, no red shift and rather, slightly blue shift was observed that PL peak position of NBE transition was 379.1 nm (3.270 eV) and 377.3 nm (3.285 eV) in undoped ZnO and ZnO:F(0.7 at.%) films, respectively. This result are well matched previous reports concerning with PL analysis in ZnO:F films [113, 114]. This blue shift with an increase of carrier concentration in ZnO:F film is opposite tendency with the red shift in NBE transition region of cation-doped ZnO films [109, 112]. This opposite tendency can be interpreted that the band gap narrowing phenomena from the many-body effect increase due to the increase of perturbation effect in conduction band by higher carrier concentration trough cation-doping in ZnO. The conduction band of ZnO are composed with the empty 4s levels of Zn^{2+} ions (from ionic characteristics) or with the lowest anti-bonding sp^3 hybrid states (from covalent characteristics). In the same manner, the valence band of ZnO originates from the filled 2p levels of O^{2-} ions or the energetically highest bonding sp^3 states. Therefore, F doping behavior of substitution for O, rather than Zn, would relatively prevent the perturbation in the conduction band compared with metal dopants, which substitute to Zn. This effect leads to significantly decrease the many-body effect in conduction band for ZnO:F films even though the addition of F^- anions raises the Fermi level of ZnO:F films into the conduction band due to filled unoccupied states in conduction bands in the most heavily doped samples (ZnO:F(0.7 at.%); low 10^{20} cm^{-3}). In other words, in the ZnO:F films, the perturbation effect in conduction band is not significant compared with cation-doped ZnO because of the different doping mechanism as mentioned in Figure 19. On the other hands, in the region of 500 nm PL spectra, the broad peak is generally explained the emission derived by the oxygen defect structure. Therefore, the significant diminishing of the intensity of visible emission (~500 nm) can be resulted by the decrease of O vacancies, which are attributable the F ions to occupy the empty sites effectively in the ZnO matrix [113, 114]. Therefore, the high m^* value in Figure 20b might be dominantly caused by the quantum confinement effect [103] due to very small grain sizes of nanocrystalline ZnO and ZnO:F films (Table 4) and also, nonparabolicity of conduction band rather than the band gap shrinkage from BGR comparing with cation-doped ZnO films [100, 103, 111]. Thus, the band gap shrinkage could be negligible in ZnO:F films because the band edge shift was not observed in NBE transition. For the simulation of electronic band structure of the films, the relative difference in Fermi energy level between undoped ZnO and ZnO:F films should be defined clearly.

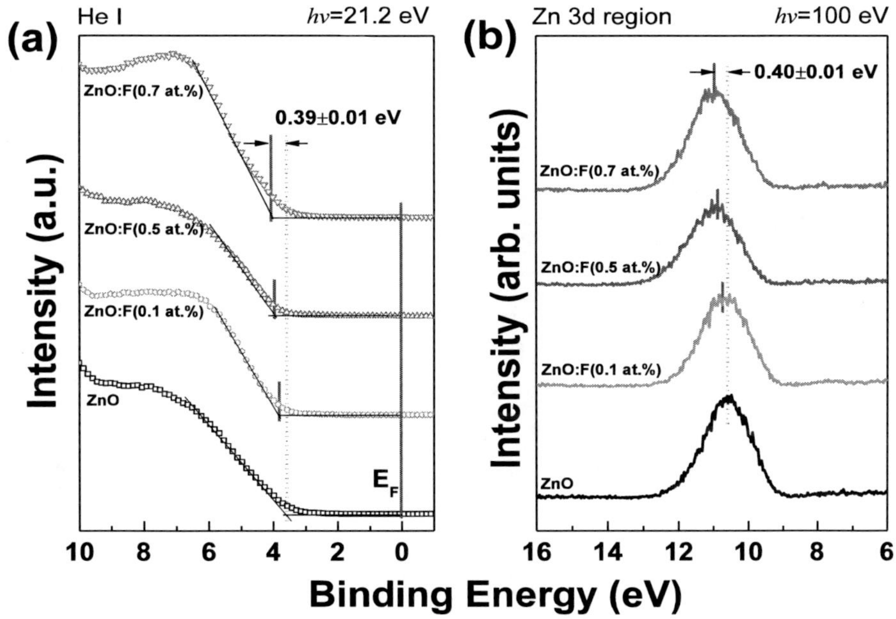

Figure 22. (a) The valence band spectra and (b) PES spectra in the Zn 3d region of undoped ZnO and various contents of ZnO:F thin films [105].

Figure 22a and b show the valence band spectra (He I radiation) and PES Zn 3d spectra for ZnO and ZnO:F thin films on the SiO_2(100 nm)/Si substrate with different electron concentration. The main features that can be identified correspond to the highest ZnO valence band, which is mainly derived from the O 2p orbitals and the Zn 3d band [96, 110]. These bands appear to be located at 3.60 eV and 10.57 eV from the E_F in a nondegenerate ZnO sample, respectively. As the ZnO conduction band populates, these features appear progressively shift to higher binding energies. In ZnO:F(0.7 at.%) films, the valence band spectra (Figure 22a) and Zn 3d PES spectra were shifted by 0.39±0.01 eV and 0.40±0.01 eV higher than those of undoped ZnO films, respectively. From this energy shift, the Fermi energy shift between undoped ZnO and ZnO:F(0.7 at.%) films by band filling could be confirmed by ~0.40 eV.

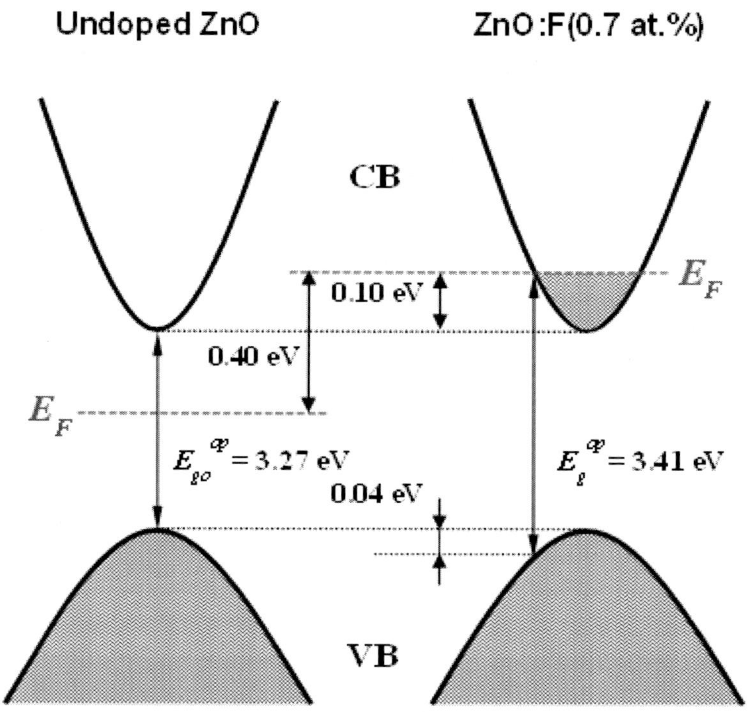

Figure 23. Summary of the band gap widening and the band gap narrowing phenomena contributions to the ΔE_g^{op} obtained experimentally for undoped ZnO and ZnO:F (0.7 at.%) thin films with $n = 1.524 \times 10^{19}$ cm^{-3} and 1.419×10^{20} cm^{-3}, respectively [105].

Schematic band diagram of ZnO and ZnO:F (0.7 at.%) films are given in Figure 23. In this figure, Fermi level of approximately 0.3 eV was estimated by UPS result and a literature. As discussed, we can define the individual energy fraction of band gap widening in the valence and conduction band of undoped ZnO and ZnO:F (0.7 at.%) films through the combination of Figures 20-22. From this result, the contribution of bandgap widening and narrowing in the ZnO:F film can be calculated by 100 meV and 40 meV, respectively. As shown in this band structure, the optoelectronic properties of ZnO films with doping of F can be analyzed with theoretical and experimental results.

2-3. Effect of Surface-Terminated Bonding Changes on the Work Function Variation of F-Doped ZnO Films

In this sub chapter, we studied atomic layer deposited ZnO and ZnO:F films at a low temperature of 140°C for application in OLED devices in order to ascertain their performance as TCO anodes. These films contained drastically textured morphologies with an increase of F concentration in ZnO matrix. In order to investigate the effects of the morphological changes on the operating performance of OLED devices, 250 to 300-nm-thick ZnO and ZnO:F anodes with F doping concentrations of 0.2, 0.5, and 0.7 at.% were grown at 140°C to control the work function of the anode. This concept of the only surface texturing for control of the work function without a buffer layer between the anode and organic layer originated from the previous observation on the surface passivation effect with F doping in ZnO films. The surface texturing on the ZnO:F anodes, which has a higher work function than that of the ZnO anode, is expected to play the role of a transition metal oxide such as MoO_3, V_2O_5, NiO, and WO_3, in order to improve the OLED device performance through effective hole injection and transport from the ZnO:Al anode to the organic layer without any kind of additional processing to insert a buffer layer. Moreover, in this investigation, the ZnO:F anodes can be formed by using a novel home-made F source and their work function can be controlled by surface texturing resulted in F doping concentration to improve the interface properties between the TCO anode and the organic layer.

2-3-1. Experimental Details

The prepared ZnO:F films were prepared on glass (Fusion 1737) substrates via ALD using a traveling wave type Lucida D100 system (NCD Technology, Inc. Korea) at a deposition temperature of 140°C and a working pressure of 3 to 3.3 Torr. Diethylzinc (DEZ, Hansol Chemical Co., Ltd., Korea) and DI water were used as a Zn precursor and a reactant source for O, respectively. F-doping into ZnO was performed by using a mixture of DI water 50 mL and diluted hydrogen fluoride (HF 48~51% diluted in water) 0.5 mL as a single reactant source for oxygen and the F dopant. For the control of F doping concentration in ZnO thin films, the cycle ratio of DEZ/H_2O to DEZ/H_2O:HF was controlled 4:1, 1:1, and 1:2. The concentration of F in ZnO in terms of atomic percent could be determined with 0.2 at.%, 0.5 at.%, and 0.7 at.% for 4:1, 1:1, and 1:2 samples, respectively, from quantitative analysis by using EDX. For the deposition of ZnO and ZnO:F thin films, total ALD cycle was 1000 total cycles for each samples and the deposition rate was approximately 2.1~2.3 Å for cycle in the range of reported value. The film thicknesses are determined by an ellipsometer and FE-SEM.

In order to apply the OLED devices as a TCO anode and buffer layer, the undoped ZnO and ZnO films with 0.2 at.% F, 0.5 at.% F, and 0.7 at.% F were cleaned using acetone, methanol, DI water, and isopropyl alcohol in an ultrasonic bath. The films were then patterned by conventional photolithography and etched over the 2 mm × 2 mm emitting area by a wet process using a mixed etchant solution of MA-SO_2 (HCl:HNO_3 based acid solution, Dongwoo Fine-Chem. Cop.):H_2O = 1:10 for 20 sec. The OLED devices with a structure of DNTPD/TAPC/$Bebq_2$:10% doped RP-411/Bphen/LiF/Al were fabricated onto the prepared ZnO and ZnO:F films. All organic and cathode materials were deposited by thermal evaporation under a pressure of 5×10^{-7} Torr.

DNTPD was used as a HIL and effectively reduced the LUMO energy barrier between the anode and the TAPC HTL. TAPC acted as a hole transport into the emission layer as well as an EBL with an energy gap 1.0 eV higher than the HOMO. The EML was composed of $Bebq_2$ as a host and RP-411 as a red phosphorescent dopant. Bphen was used as an ETL and a HBL, while LiF/Al was used as the EIL and cathode.

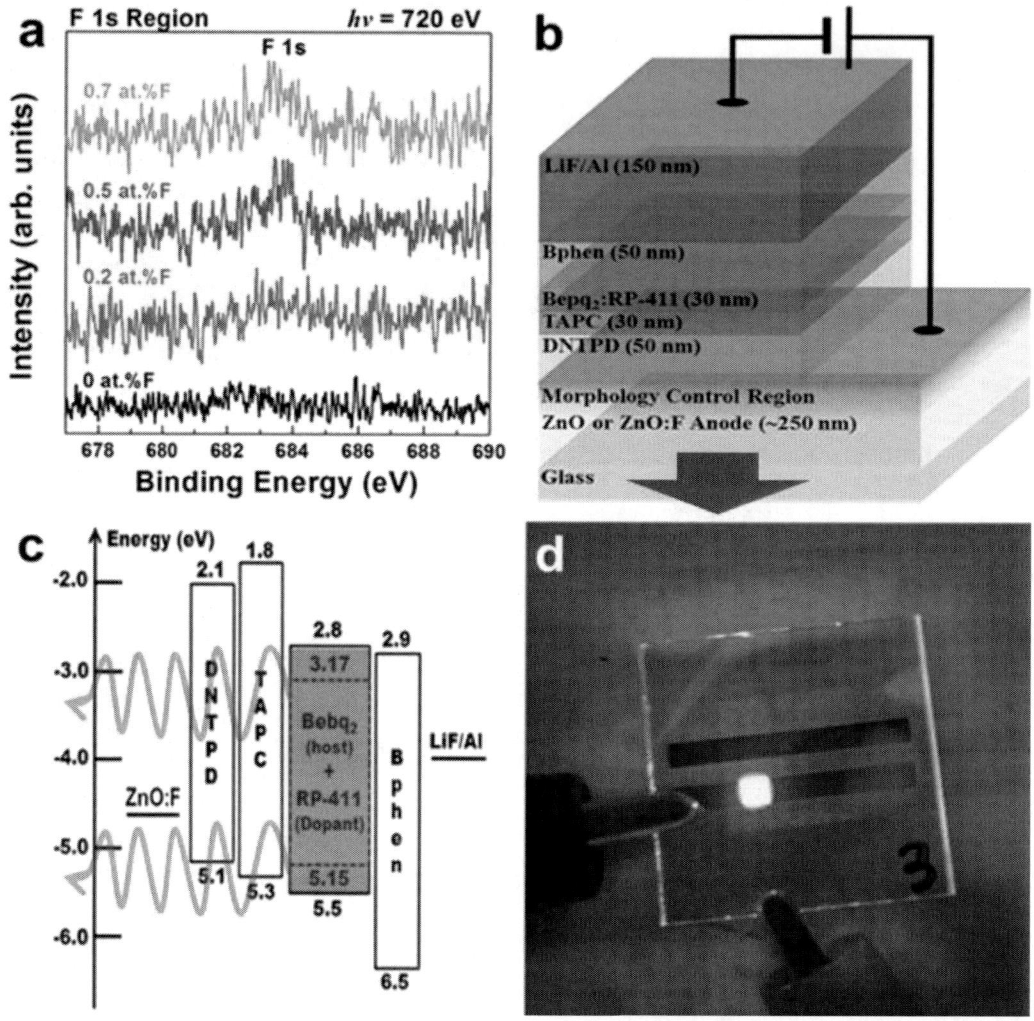

Figure 24. (a) The PES F1s region as a function of the F doping concentration, (b) the OLED device structure fabricated on ZnO-based TCO anode, (c) a schematic of the energy band diagram of the OLED device, and (d) real image of the OLED device on an ZnO:F (0.5 at.%) anode [115].

The crystal structure and surface morphology of the ZnO:F anode films were analyzed by XRD with $CuK\alpha_1$ radiation and FE-SEM. A quantitative analysis of F doping concentration was conducted using EDX. The electrical resistivity of the ZnO:F films was measured using a Hall effect measurement system (HMS-3000, Ecopia) under 10 mA and a 0.57 Tesla electromagnetic field. The transmittance of the ZnO:F films was inspected using a UV-vis-NIR spectrometer. UPS analysis was performed with He I (21.22 eV) radiation and a bias voltage (-5 V) in the 4D beam line of the Pohang Accelerator Laboratory (PLS). Au foil was used as a reference material for the calibration of the kinetic energy shift. The electrical and optical characteristics of the OLED as a function of the ZnO:F anodes were investigated using an I-L-V measurement system (M6100, McScience) in a dark room. Figure 24 shows (a) the PES F1s region as a function of the F doping concentration, (b) the OLED device structure fabricated on ZnO-based TCO anode, (c) a schematic of the energy band diagram of the OLED device, and (d) real image of the OLED device on an ZnO:F (0.5 at.%) anode.

2-3-2. Structural, Optical, and Electrical Properties of ZnO:F Films by F Doping Concentration

Figure 25 shows the XRD spectra of the films grown by 1000 total ALD cycles at a constant temperature of 140°C. The right side figure is indicated by enlarged 2 theta region from 31.5° to 36° in Figure 25. The main peaks of all the films at 2 theta of 31.73°, 34.46°, and 56.59° were indicated by (100), (002), and (110) plane, respectively. Wurtzitie ZnO without observation of any lattice distortion with F doping into ZnO matrix was due to similar ionic size of O^{2-} and F^{-} [65]. The optical transmittance curves ranging from 200 nm to 1000 nm of undoped and doped ZnO films are shown in Figure 26. The red region in Figure 26 is corresponded with the red light wavelength ranging from 625 to 740 nm, which is the emission light wavelength from OLED operation in this work and also insetimage was presented by the patterned TCO anode before a deposition of the organic layers. The average transmittances values of all the films were about 85% in the visible region (380 to 780 nm). Figure 27 shows the sheet resistance and the average transmittance values in the red light region of undoped ZnO and various ZnO:F films. The average transmittance values were very slightly increased by an increase of F concentration and those of films were over 80% for all the films. The sheet resistance of undoped ZnO, ZnO:F (0.2 at.%), ZnO:F(0.5 at.%), and ZnO:F (0.7 at.%) films were 1500, 197, 168, and 125 Ω sq^{-1}, respectively.

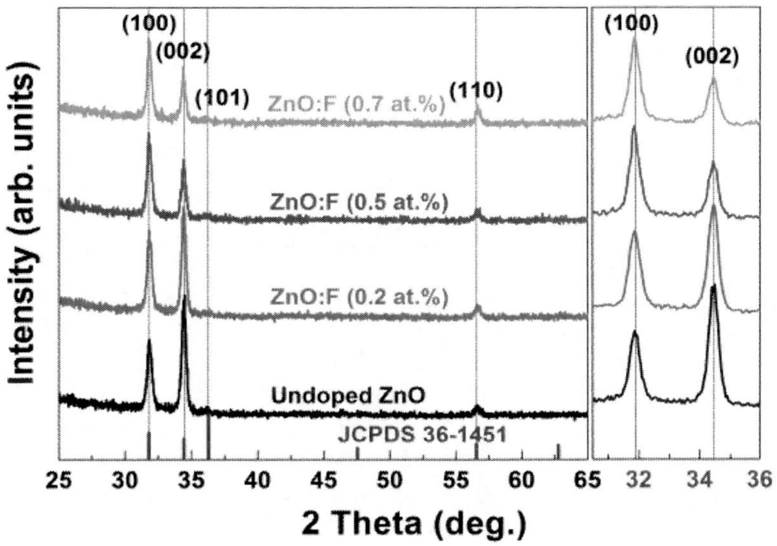

Figure 25. XRD spectra of the films grown by 1000 total ALD cycles at a constant temperature of 140°C and the right side figure is indicated by enlarged 2 theta region from 31.5° to 36° [115].

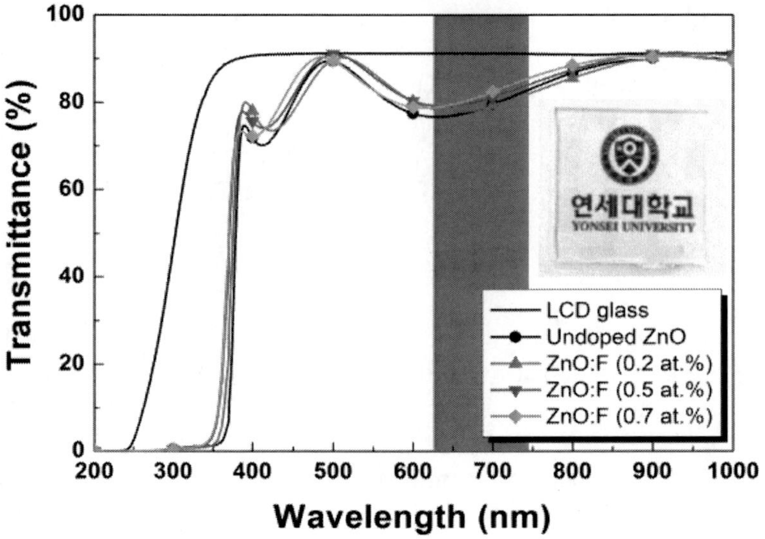

Figure 26. The optical transmittance curves ranging from 200 nm to 1000 nm of undoped and doped ZnO films. The red region is corresponded with the red light wavelength ranging from 625 to 740 nm, which is the emission light wavelength from OLED operation in this work and also inset image is presented by the patterned ZnO:F anode before a deposition of the organic layers [115].

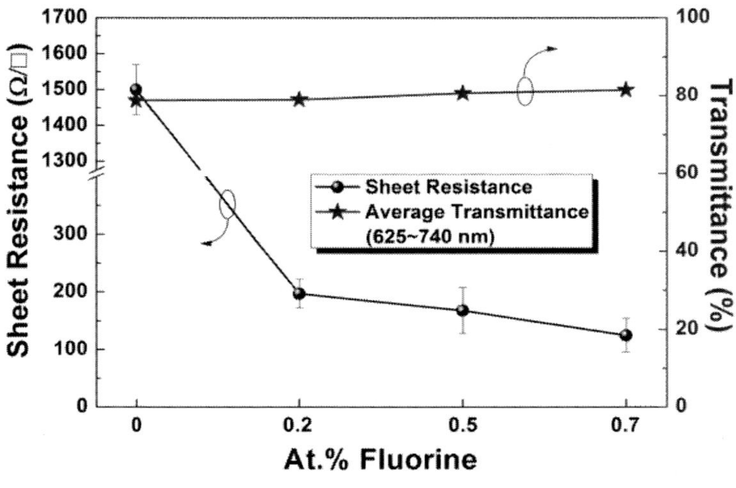

Figure 27. The sheet resistance and the average transmittance values in the red light region of undoped ZnO and various ZnO:F films as a function of F doping concentration [115].

The cross-sectional FE-SEM images and corresponding AFM images and RMS roughness values of the (a) undoped ZnO, (b) ZnO:F (0.2 at.%), (c) ZnO:F (0.5 at.%), and (d) ZnO:F (0.7 at.%) films are shown in Figure 28. RMS roughness values were monotonically increased by an increase of the F doping concentration as well matched with the reported phenomenon in previous section. These morphological changes with F doping concentration might be influenced by F doping mechanism in ZnO matrix, indicating the F^- anions filled with O- related defect site or substitution in O^{2-} sites. And also, surface polarity of ZnO and ZnO:F films could be changed by this morphological changes in the surface due to the change of the fraction of the surface-exposed planes. And also, the peaks are agreed well with the standard diffraction pattern (JCPDS 36-1451).

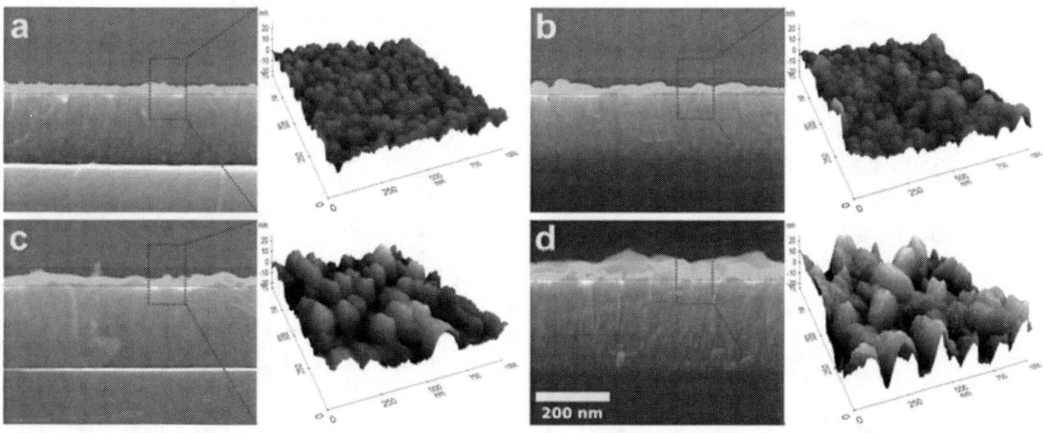

Figure 28. The cross-sectional FE-SEM images and corresponding AFM images and RMS roughness values of the (a) undoped ZnO, (b) ZnO:F (0.2 at.%), (c) ZnO:F (0.5 at.%), and (d) ZnO:F (0.7 at.%) films [115].

2-3-3. Work Function Variation of ZnO:F Films by Surface-Terminated Bonding Changes by F Doping Concentration

Figure 29 shows (a) UPS spectra of the low kinetic energy cutoff region and the valence band regions of ZnO and various contents of ZnO:F films, and (b) the schematic diagram of the surface of ZnO and ZnO:F films with expected surface planes. UPS spectra for the secondary cutoff region and valance band spectra region were collected using a He I (hυ = 21.22eV) energy source. The spectrum was taken with a -5.0 V sample bias so that the sample inelastic cutoff could be distinguished from that due to the spectrometer in Figure 29. The Fermi energy (E_F) of the films was 21.35 eV, as indicated by the onset of energy intensity in Figure 29. The work function of the films was determined using Eq. 18: [116]

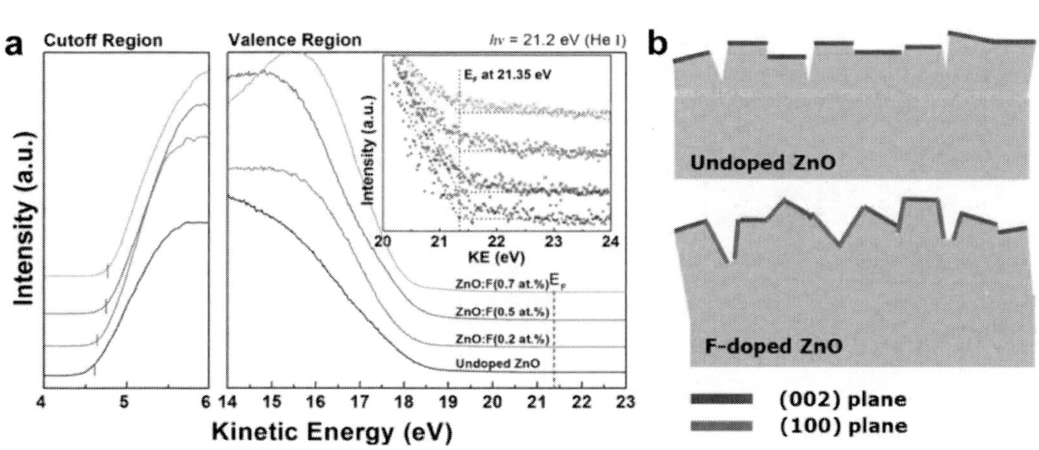

Figure 29. (a) UPS spectra of the low kinetic energy cutoff region and the valence band regions of ZnO and various contents of ZnO:F films, and (b) the schematic diagram of the surface of ZnO and ZnO:F films with expected surface planes [115].

$$\Phi = h\upsilon + E_{cutoff} - E_F \quad \text{(Eq. 18)}$$

From Eq. 18, the work functions of undoped ZnO, ZnO:F (0.2 at.%), ZnO:F (0.5 at.%), and ZnO:F (0.7 at.%) films were 4.60, 4.63, 4.74, and 4.76 eV, respectively. The work function values of the ZnO:F films clearly increased with an increase in F doping concentration as well as with the increase of RMS roughness values. An increase in the F content lead to the textured film surface and finally, the portion of (100) plane per (002) plane might be increased by the exposed (100) planes. This change of surface polarity of the films allows the increase of ionization potential of the films with an increase of the F doping concentration, resulting in the increase of work function of TCO films.

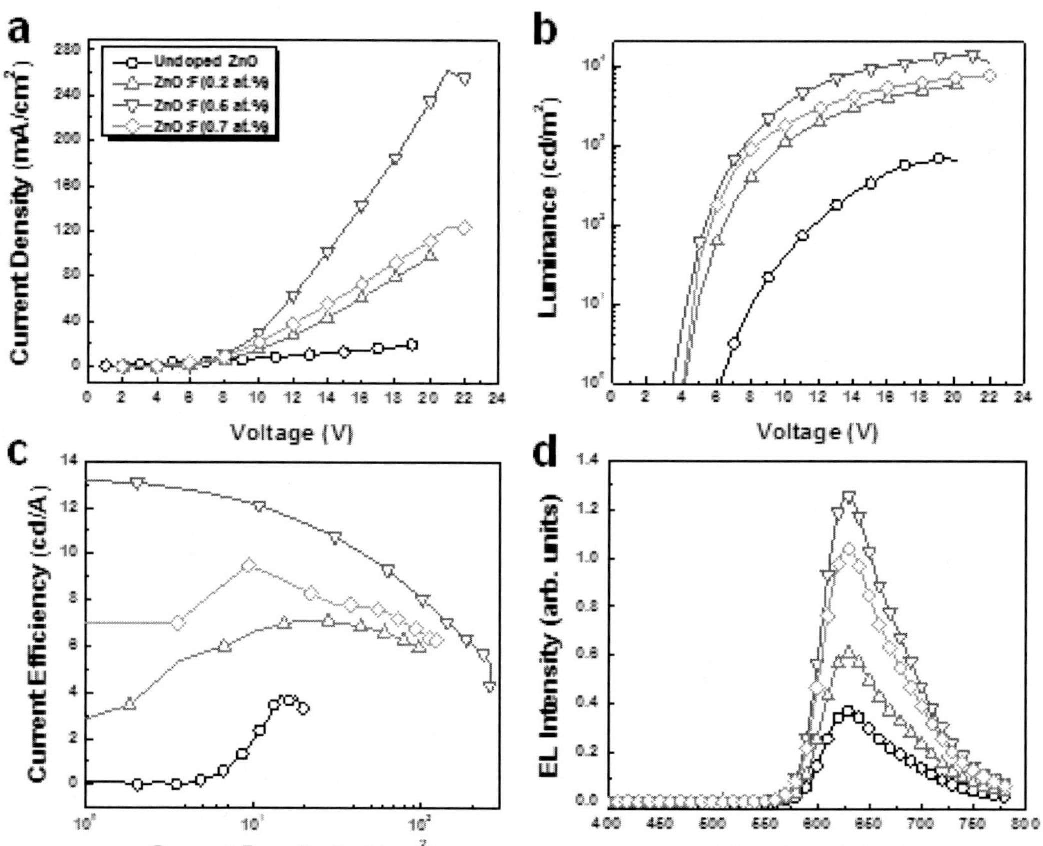

Figure 30. (a) Current density versus voltage (J-V), (b) luminance versus voltage (L-V), and (c) current efficiency versus current density (Ec-J) characteristics and (d) EL efficiency of the OLED devices fabricated onto undoped ZnO and various contents of ZnO:F films [115].

2-3-4. Effect of the Surface-Terminated Bonding Changes of ZnO:F Films on OLED Performances as a Transparent Anode

Figure 30 shows (a) Current density versus voltage (J-V), (b) luminance versus voltage (L-V), and (c) current efficiency versus current density (Ec-J) characteristics and (d) EL efficiency of the OLED devices fabricated onto undoped ZnO and various contents of ZnO:F films. The current density values at 20 V of OLED devices fabricated onto ZnO and doped 247.98, and 110.31 mA/cm^2 for undoped ZnO, ZnO:F (0.2 at.%), ZnO:F (0.5 at.%), and ZnO:F (0.7 at.%), respectively. The luminance values of all prepared OLED devices onto undoped and doped ZnO films were presented by similar tendency with J-V characteristics. The maximum luminance values of OLED devices were 697.73, 5705.60, 13509.55, and 7982.97 cd/m^2 for undoped ZnO, ZnO:F (0.2 at.%), ZnO:F (0.5 at.%), and ZnO:F (0.7 at.%), respectively, at a bias of 20 V. The turn-on voltage gap of OLED devices of the samples becomes more important at higher luminance values. At a luminance of 10 cd/m^2, the turn-on voltage values of OLED devices were 8.0, 5.0, 4.1, and 4.6 V for undoped ZnO, ZnO:F (0.2 at.%), ZnO:F (0.5 at.%), and ZnO:F (0.7 at.%), respectively. At a luminance of 1000 cd/m^2, the turn-on voltage values were N/A, 9.7, 7.5, and 8.2 V, respectively. These results were due to combination of a decrease in the sheet resistance and an increase in the work function of the anode. The highest performances were shown in the OLED devices onto ZnO:F (0.5 at.%) anode. This might be influenced by an drastic increase of RMS roughness values for ZnO:F (0.7 at.%) sample, indicating 3 times more than that of ZnO:F (0.5 at.%) samples. Therefore, the RMS roughness values might act as critical degradation factors at the interface between organic layer, because ZnO:F (0.5 at.%) film has the similar sheet resistance and work function values with those of ZnO:F (0.7 at.%) film. Similar to the result shown in Figure 30a and b, higher current efficiency and EL efficiency of the OLED devices were obtained for ZnO:F (0.5 at.%) as shown in Figure 30c and d.

CONCLUSION

In this chapter, a simple method of F doping into a ZnO matrix using ALD was mainly focused. The deposition temperature of the ALD-ZnO:F films was very low (140°C), yet still yielded excellent electrical and optical performances compared with other reported methods that require growth temperatures over 450°C. And also, extrinsic F doping mechanism was verified by detailed analysis of the growth mode of ZnO:F films through the corroboration by the XRD, surface morphology, and GIWAXD analysis. This F doping phenomenon could be explained by the substitution of O sites with F anions, resulting in the defect structures of ZnO films filled with F anions. Therefore, along with the increase in the F doping concentration, the optical transmittance of the films increased slightly from 80.97% to 83.02% due to a decrease in the absorption in the visible region by F anions filling O defects. The Burstein-Moss effect may also explain to the higher transparency of the films through the band gap widening effect with the increase in carrier concentration induced by the increase in F doping concentration. The electron carrier concentration was gradually increased from 1.713×10^{19} cm^{-3} to 1.462×10^{20} cm^{-3} by the increase in the F concentration in the films. The Hall mobility was maximized at 43.66 cm^2 V^{-1} s^{-1} with a low concentration of F doping of approximately 0.20 at.%. A minimum resistivity value of 1.876×10^{-3} Ω cm was obtained in ZnO:F(1.0) films, and the carrier concentration and mobility were 1.375×10^{20} cm^{-3} and 24.20 cm^2 V^{-1} s^{-1}, respectively. Therefore, it was primarily confirmed that the novel fabrication method of ZnO:F thin films using ALD and an adjusted F doping concentration could be optimized for a growth temperature of 140°C, while very little work has been done on the ZnO:F films comparing with cation-doped ZnO films.

The electron band structure of ZnO:F films was discussed with the band-gap widening and band-gap narrowing effect on them. The increase of electron concentration by F doping within ZnO matrix resulted in optical-band-gap widening and slight narrowing, explaining by band-filling model and BGR effect. And also, their individual contribution to the band gap shift could be calculated by the observation of the valence band spectra, UPS measurement, absorption edge shift, and PL-NBE transition. This result provides the information of the band structure of anion-doped ZnO films comparing with cation-doped ZnO films.

The OLED devices were tested by using undoped ZnO anode and ZnO films doped with 0.2 at.%, 0.5 at.%, and 0.7 at.% F. As a result, the OLED device performance in terms of luminance, current efficiency, and turn-on voltage was improved by controlling the surface texturing of undoped ZnO films by doping with F, which has the doping mechanism with passivating the ZnO surface as demonstrated in this chapter. This morphological change resulted in the variation of work function ZnO anode, and the increase of work function allowed the increase of effectiveness of hole injection and transport between the anode and DNTPD HIL. Therefore, the controllability of surface texturing and morphology by F doping in ZnO films can improve the characteristics of OLED devices by controlling the nature of the interface in terms of sheet resistance, work function, and RMS roughness. Moreover, these fluorinated anode materials can also be applied to any kind of OLED structure.

In summary, halogen doping method by using home-made halogen sources for use in ALD was mainly demonstrated and the doping mechanism of halogen dopants in ZnO matrix discussed in detail. The results and interpretations provided in this chapter are desired to give the guideline and information for halogen doping in oxide materials fabricated by ALD.

REFERENCES

[1] S. Bae, H. Kim, Y. Lee, X. Xu, J.-S. Park, Y. Zheng, J. Balakrishnan, T. Lei, H.R. Kim, Y.I. Song, Y.-J. Kim, K.S. Kim, B. Özyilmaz, J.-H. Ahn, B.H. Hong, S. Iijima, *Nature 5*, 574 (2010).

[2] M. Zhang, S. Fang, A.A. Zakhidov, S.B. Lee, A.E. Aliev, C.D. Williams, K.R. Atkinson, R.H. Baughman, *Science 309*, 1215 (2005).

[3] L. Hu, H.S. Kim, J.-Y. Lee, P. Peumans, Y. Cu, *ACS Nano 4*, 2955 (2010).

[4] D.S. Ginley, C. Bright, *MRS Bull. 25*, 15 (2000).

[5] E. Fortynato, D. Ginley, H. Hosono, D.C. David, *MRS Bull. 32*, 242 (2007).

[6] Y.-J. Choi, S.C. Gong, C.-S. Park, H.-S. Lee, J.G. Jang, H.J. Chang, G.Y. Yeom, H.-H. Park, *ACS Appl. Mater. Interfaces 5*, 3650 (2013).

[7] K. Nomura, H. Ohta, A. Takagi, T. Kamiya, M. Hirano, H. Hosono, *Nature 432*, 48 (2004).

[8] T. Minami, *Semicond. Sci. Technol. 20*, S35 (2005).

[9] R.B. Tahar, T. Ban, Y. Ohya, Y. Takahashi, *J. Appl. Phys. 82*, 865 (1997).

[10] K. Ellmer, *J. Phys. D: Appl. Phys. 34*, 3097 (2001).

[11] H. Hosono, *J. Non-Cryst. Solids 352*, 851 (2006).

[12] T. Minami, *MRS Bulletin 25*, 38 (2000).

[13] M.L. Olvera, A. Maldonado, R. Asomoza, *Sol. Energy Mater. Sol. Cells 73*, 425 (2002).

[14] J. Hu, R.G. Gordon, *Sol. Energy Mater. Sol. Cells 30*, 437 (1991).

[15] H.Y. Xu, Y.C. Liu, R. Mu, C.L. Shao, Y.M. Lu, D.Z. Shen, X.W. Fan, *Appl. Phys. Lett. 86*, 123107 (2005).

[16] L. Cao, L. Zhu, J. Jiang, R. Zhao, Z. Ye, B. Zhao, *Sol. Energy Mater. Sol. Cells 95*, 894 (2011).

[17] H.S. Yoon, K.S. Lee, B. Cheong, D.K. Choi, D.H. Kim, W.M. Kim, *Sol. Energy Mater. Sol. Cells 92*, 1366 (2008).

[18] K. Badeker, *Ann. Phys. (Leipzig) 22*, 749 (1907).

[19] R.M. Pasquarelli, D.S. Ginley, R. O'Hayre, *Chem. Soc. Rev. 40*, 5406 (2011).

[20] J.F. Wager, D.A. Keszler, R.E. Presley, "Transparent Electronics," Springer, New York, USA, (2008).

[21] A. Facchetti, T.J. Marks, "Tranparent Electronics from Synthesis to Applications," Wiley, United Kindom, (2010).

[22] A.J. Freeman, K.R. Poeppelmeier, T.O. Mason, R.P.H. Chang, T.J. Marks, *MRS Bulletin 25*, 45 (2000).

[23] J.E. Medvedeva, A.J. Freeman, *Europhys. Lett. 69*, 583 (2005).

[24] E. Burstein, *Phys. Rev. 93*, 632 (1954).

[25] T.S. Moss, *Proc. Phys. Soc. B 67*, 775 (1954).

[26] R.G. Gordon, *MRS Bull 25*, 52 (2000).

[27] H. Morkoç, Ü. Özgür, "Zinc Oxide: Fundametals, Materials and Device Technology," Wiley-VCH, Weinheim, (2009).

[28] Ü. Özgür, Y. I. Alivov, C. Liu, A. Teke, M.A. Reshchikov, S. Doğan, V. Avrutin, S.-J. Cho, H. Morkoç, *J. Appl. Phys. 98*, 041301 (2005).

[29] T. Wang and P. V. Radovanovic, *J. Phys. Chem. C 115*, 406 (2011).

[30] G.B. Palmer, K.R. Poeppelmeier, T.O. Mason, *Chem. Mater. 9*, 3121 (1997).

[31] Y. Kim, W. Lee, D.-R. Jung, J. Kim, S. Nam, H. Kim, B. Park, *Appl. Phys. Lett. 96*, 171902 (2010).

[32] S.T. Tan, B.J. Chen, X.W. Sun, W.J. Fan, H.S. Kwok, X.H. Zhang, S.J. Chua, *J. Appl. Phys. 98*, 013505 (2005).

[33] L. Cao, L. Zhu, J. Jiang, R. Zhao, Z. Ye, B. Zhao, *Sol. Energy Mater. Sol. Cells 95*, 894 (2011).

[34] J.A. Sans, J.F. Sánchez-Royo, A. Segura, G. Tobias and E. Canadell, *Phys. Rev. B 79*, 195105 (2009).

[35] H.Q. Le, S.J. Lim, G.K.L. Goh, S.J. Chua, J. Ong, *J. Electrochem. Soc. 157*, H196 (2010).

[36] C.F. Klingshirn, B.K. Meyer, A. Waag, A. Hoffmann, J. Geurts, "Zinc Oxide from Fundamental Properties towards Novel Applications," Spirnger, Newyork, USA, (2010).

[37] *NPG Asia Materials* research highlight; doi:10.1038/asiamat.2009.216, Published online 15 June 2009.

[38] U. Ozgur, H. Morkoc, *Journal of Applied Physics 98*, 041301 (2005).

[39] J. L. Vossen, *Phys. Thin Films 9*, 1 (1977).

[40] R.J. Lad, P.D. Funkenbusch, C.R. Aita, *J. Vac. Sci. Technol 17*, 808 (1980).

[41] M.K. Ryu, S.H. Lee, M.S. Jang, G.N. Panin, T.W. Kang, *J. Appl. Phys 92*, 154 (2002).

[42] P. Fons, K. Iwata, S. Niki, A. Yamada, K. Matsubara, M. Watanabe, *J. Cryst. Growth 209*, 532 (2000).

[43] D.B. Christey, G.K. Hubler, "Pulsed laser deposition of thin films," John Wiley & Sons Inc. (1994).
[44] Y. R. Ryu et al., *Appl. Phys. Lett. 88*, 201 (2000).
[45] A. Tsukazaki et al., *Nat. Mater. 4*, 42 (2005).
[46] C. Jagadish, S. Pearton, "Zinc oxide bulk, thin films and nanostructures," Elsevier, Oxford, UK (2006).
[47] C.K. Lau, S.K. Tiku, K.M. Lakin, *J. Electrochem. Soc. 127*, 1843 (1980).
[48] F.T.J. Smith, *Appl. Phys. Lett. 43*, 1108 (1983).
[49] Y. Kashiwaba, K. Haga, H. Watanabe, B. P. Zhang, Y. Segawa, K. Wakatsuki, *Phys. Status Solidi B 229*, 921 (2002).
[50] T. Yasuda, Y. Segawa, *Phys. Status Solidi B 241*, 676 (2004).
[51] N. Oleynik, *J. Cryst. Growth 248*, 14 (2003).
[52] T. Kaufmann, G. Fuchs, W. Webert, *Cryst. Res. Technol. 23*, 635 (1988).
[53] W. Kern, R.C. Heim, *J. Electrochem. Soc. 117*, 562 (1970).
[54] A. Dadgar, *J. Cryst. Growth 267*, 140 (2004).
[55] T. Yasuda, Y. Segawa, *Phys. Status Solidi B 241*, 676 (2004).
[56] J.J. Robbins, James Esteban, Cassandra Fry, C.A. Wolden, *J. Electrochemical society 150,* C693 (2003).
[57] B.S. Li, Y.C. Liu, D.Z. Shen, Y.M. Lu, J.Y. Zhang, X.G. Kong, X.W. Fan, *J. Vac. Sci. Technol. A 20*, 265 (2002).
[58] T. Suntola, J. Antson, U.S. Patent No. 4,058,430 (1977).
[59] M. Ritala, M. Leskelä, in *Handbook of Thin Film Materials*, edited by H.S. Nalwa Academic, San Diego, 2002, Vol. 1, pp. 103–159.
[60] J.L. Stickney, *Electroanal. Chem. 21*, 75 (1999).
[61] S. Lindroos, T. Kanniainen, M. Leskelä, *Appl. Surf. Sci. 75*, 70 (1994).
[62] R.L. Puurunen, *J. Appl. Phys. 97*, 121301 (2005).
[63] S.M. George, A.W. Ott, J.W. Klaus, *J. Phys. Chem. 100*, 13121 (1996).
[64] A.M. Hoyas, J. Schuhmacher, D. Shamiryan, J. Waeterloos, W. Besling, J.P. Celis, K. Maex, *J. Appl. Phys. 95*, 381 (2004).
[65] Y.-J. Choi and H.-H. Park, *J. Mater. Chem. C 2*, 98 (2014)
[66] J.-S. Na, Q. Peng, G. Scarel, G. N. Parson, *Chem. Mater. 21*, 5585 (2009).
[67] J.W. Elam, S.M. George, *Chem. Mater. 15*, 1020 (2003).
[68] J. Hu, R.G. Gordon, *Sol. Energy Mater. Sol. Cells 30*, 437 (1991).
[69] H. Y. Xu, Y. C. Liu, R. Mu, C.L. Shao, Y.M. Lu, D.Z. Shen, X.W. Fan, *Appl. Phys. Lett. 86*, 123107 (2005).
[108] H.S. Yoon, K.S. Lee, B. Cheong, D.K. Choi, D.H. Kim, W.M. Kim, *Sol. Energy Mater. Sol. Cells 92*, 1366 (2008).
[70] P. Banerjee, W.-J. Lee, K.-R. Bae, S.B. Lee, G.W. Rubloff, *J. Appl. Phys. 108*, 043504 (2010).
[71] A. Segura, J.A. Sans, D. Errandonea, D. Martinez-García, V. Fages, *Appl. Phys. Lett. 88*, 011910 (2006).
[72] B. Liu, M. Gu, X. Liu, S. Huang, C. Ni, *Appl. Phys. Lett. 97*, 122101 (2010).
[73] D.-J. Lee, H.-M. Kim, J.-Y. Kwon, H. Choi, S.-H. Kim, K.-B. Kim, *Adv. Funct. Mater. 21*, 448 (2011).
[74] S.-Y. Pung, K.-L. Choy, X. Hou, C. Shan, *Nanotechnology 19*, 435609 (2008).

[75] L. Cao, L. Zhu, J. Jiang, R. Zhao, Z. Ye, B. Zhao, *Sol. Energy Mater. Sol. Cells 95*, 894 (2011).
[76] S.-H. Ko Park, Y.E. Lee, *J. Mater. Sci. 39*, 2195 (2004).
[77] Berginski, J. Hüpkes, M. Schulte, G. Schöpe, H. Stiebig, B. Rech, *J. Appl. Phys. 101*, 074903 (2007).
[78] K. Vanheusden, W.L. Warren, C.H. Seager, D.R. Tallant, J.A. Voigt, B.E. Gnade, *J. Appl. Phys. 79*, 7983 (1996).
[79] Y.-P. Wang, W.-I. Lee, T.-Y. Tseng, *Appl. Phys. Lett. 69*, 1807 (1996).
[80] Y.-Z. Tsai, N.-F. Wang, C.-L. Tsai, *Mater. Lett. 63*, 1621 (2009).
[81] H.Y. Xu, Y.C. Liu, J.G. Ma, Y.M. Luo, Y.M. Lu, D.Z. Shen, J.Y. Zhang, X.W. Gan, R. Mu, *J. Phys.-Condes. Matter 16*, 5143 (2004).
[82] T. Wang, P.V. Radovanovic, *J. Phys. Chem. C 115*, 406 (2011).
[83] R.G. Gordon, *MRS Bull. 25*, 52 (2000).
[84] S.T. Tan, B.J. Chen, X.W. Sun, W.J. Fan, H.S. Kwok, X.H. Zhang, S.J. Chua, *J. Appl. Phys. 98*, 013505 (2005).
[85] E. Burstein, *Phys. Rev. Lett. 93*, 632 (1954).
[86] T.S. Moss, Proc. Phys. Soc. London, Sect. B *67*, 775 (1954).
[87] H.Q. Le, S.J. Lim, G.K.L. Goh, S.J. Chua, J. Ong, *J. Electrochem. Soc. 157*, H196 (2010).
[88] J.A. Sans, J.F. Sánchez-Royo, A. Segura, G. Tobias, E. Canadell, *Phys. Rev. B 79*, 195105 (2009).
[89] B.E. Sernelius, K.-F. Berggren, Z.-C. Jin, I. Hamberg, C.G. Granqvist, *Phys. Rev. B 37*, 10244 (1988).
[90] J.D. Ye, S.L. Gu, S.M. Zhu, S.M. Liu, Y.D. Zheng, R. Zheng, Y. Shi, *Appl. Phys. Lett. 86*, 192111 (2005).
[91] H. Saarenpää, T. Niemi, A. Tukiainen, H. Lemmetyinen, N. Tkachenko, *Sol. Energy Mater. Sol. Cells 94*, 1379 (2010).
[92] A.B.F. Martinson, J.W. Elam, J.T. Hupp, M. J. Pellin, *Nano Lett. 8*, 2183 (2007).
[93] P.F. Carcia, R.S. McLean, M.H. Reilly, *Appl. Phys. Lett. 88*, 123509 (2006).
[94] J. Meyer, P. Görrn, S. Hamwi, H.-H. Johannes, T. Riedl, W. Kowalsky, *Appl. Phys. Lett. 93*, 073308 (2008).
[95] S.-H. K. Park, C.-S. Hwang, M. Ryu, S. Yang, C. Byun, J. Shin, J.-I. Lee, K. Lee, M.S. Oh, S. Im, *Adv. Mater. 21*, 678 (2009).
[96] Y. Kim, W. Lee, D.-R. Jung, J. Kim, S. Nam, H. Kim, B. Park, *Appl. Phys. Lett. 96*, 171902 (2010).
[97] L.M. Wong, S.Y. Chiam, J.Q. Huang, S.J. Wang, J.S. Pan, W.K. Chim, *Appl. Phys. Lett. 98*, 022106 (2011).
[98] X.G. Xu, H.L. Yang, D.L. Zhang, S.Z. Wu, J. Miao, Y. Jiang, X.B. Qin, X.Z. Cao, B.Y. Wang, *Appl. Phys. Lett. 97*, 232502 (2010).
[99] M.-C. Chen, T.-C. Chang, C.-T. Tsai, S.-Y. Huang, S.-C. Chen, C.-W. Hu, S.M. Sze, M.-J. Tsai, *Appl. Phys. Lett. 96*, 262110 (2010).
[100] A.P. Roth, J.B. Webb, D.F. Williams, *Phys. Rev. B 25*, 7836 (1982).
[101] M.L. Olvera, A. Maldonado, R. Asomoza, *Sol. Energy Mater. Sol. Cells 73*, 425 (2002).
[102] H.S. Yoon, K.S. Lee, B. Cheong, D.K. Choi, D.H. Kim, W.M. Kim, *Sol. Energy Mater. Sol. Cells 92*, 1366 (2008).

[103] H.Y. Xu, Y.C. Liu, R. Mu, C.L. Shao, Y.M. Lu, D.Z. Shen, X.W. Fan, *Appl. Phys. Lett. 86*, 123107 (2005).
[104] L. Cao, L. Zhu, J. Jiang, R. Zhao, Z. Ye, B. Zhao, *Sol. Energy Mater. Sol. Cells 95*, 894 (2011).
[105] Y.-J.Choi, K.-M. Kang, H.-H. Park, *Sol. Energy Mater. Sol. Cells 132*, 403 (2015)
[106] B. Liu, M. Gu, X. Liu, S. Huang, C. Ni, *Appl. Phys. Lett. 97*, 122101 (2010).
[107] A. Janotti, C.G.V.d. Walle, *Phys. Rev. B. 76*, 165202 (2007).
[108] S.T. Tan, B.J. Chen, X.W. Sun, W.J. Fan, H.S. Kwok, X.H. Zhang, S.J. Chua, *J. Appl. Phys. 98*, 013505 (2005).
[109] H.Q. Le, S.J. Lim, G.K.L. Goh, S.J. Chua, J. Ong, *J. Electrochem. Soc. 157*, H196 (2010).
[110] J.A. Sans, J.F. Sánchez-Royo, A. Segura, G. Tobias, E. Canadell, *Phys. Rev. B 79*, 195105 (2009).
[111] B.E. Sernelius, K.-F. Berggren, Z.-C. Jin, I. Hamberg, C.G. Granqvist, *Phys. Rev. B 37*, 10244 (1988).
[112] J.D. Ye, S.L. Gu, S.M. Zhu, S.M. Liu, Y.D. Zheng, R. Zheng, Y. Shi, *Appl. Phys. Lett. 86*, 192111 (2005).
[113] Y.-Z. Tsai, N.-F. Wang, C.-L. Tsai, *Mater. Lett. 63*, 1621 (2009).
[114] H.Y. Xu, Y.C. Liu, J.G. Ma, Y.M. Luo, Y.M. Lu, D.Z. Shen, J.Y. Zhang, X.W. Gan, R. Mu, *J. Phys.-Condes. Matter 16*, 5143 (2004).
[115] Y.-J. Choi, S. C. Gong, K.-M. Kang and H.-H. Park, *J. Mater. Chem. C 2*, 8344 (2014)
[116] D.-J. Yun, K. Hong, S.H. Kim, W.-M. Yun, J.-Y. Jang, W.-S. Kwon, C.E. Park, S.-W. Rhee, *ACS Appl. Mater. Interfaces 3*, 43 (2011).

In: Atomic Layer Deposition (ALD)
Editor: Jeannie Valdez

ISBN: 978-1-63483-869-6

Chapter 2

ATOMIC LAYER DEPOSITION OF SUB-NANO TO NANOSCALE SURFACE COATINGS FOR NEXT-GENERATION ADVANCED BATTERY SYSTEMS

Xiangbo Meng [*] ***and Jeffrey W. Elam*** [†]
Energy Systems Division, Argonne National Laboratory,
Argonne, Illinois, US

ABSTRACT

Renewable clean energies such as solar radiation and wind are emerging as the only viable options to address the daunting societal challenges of increasing energy demands and climate change. Due to their geographical limitations and intermittent operations, however, solar and wind demand electrical energy storage (EES) devices for large-scale implementation. Batteries are one of the most successful electrochemical EES devices, and can store surplus solar and wind energy during the peak generation periods for later release into the grid. In addition to supporting the grid, batteries also play a critical role in electric vehicles and portable electronics. Currently, lithium-ion batteries (LIBs) dominate consumer electronics, but various performance issues (energy density, safety, cost, and stability) inhibit their widescale deployment into transportation and smart grids. In this context, a huge research effort is being directed towards developing next-generation LIBs and new "beyond lithium" battery concepts to address these technical challenges.

Atomic layer deposition (ALD), a thin film deposition technique used in the mass production of microelectronics, has recently been identified as a promising technology to improve battery performance. By depositing ultrathin (sub-nanometer to nanometer scale) films on battery electrodes (anodes and cathodes), ALD has been shown to tailor the electrode-electrolyte interfaces and produce reliable high performance, improved stability and safety, and sustained high energy density. This work provides a comprehensive review of ALD surface coatings for battery systems. The introduction presents a general background of various battery technologies, their technical challenges, and the effects of surface coatings on battery performance. We then illustrate the mechanism of ALD and how this provides unique benefits for depositing sub-nano and nanoscale surface

[*] E-mail address: xmeng@anl.gov.
[†] E-mail address: jelam@anl.gov.

coatings. We summarize the progressing advances of ALD coatings in batteries, ranging from binary oxides, to ternary lithium-containing oxides and other non-oxides (e.g., nitrides and fluorides). To demonstrate the utility of ALD surface coatings for batteries, we highlight their application in advanced battery systems including LIBs, lithium-sulfur, lithium-oxygen, and sodium-based batteries. Finally, we provide an outlook towards future opportunities for ALD surface coatings and their eventual commercialization in battery manufacturing.

1. INTRODUCTION

Nowadays fossil fuels as our main energy supplies are quickly depleting with the release of greenhouse gases while modern society is needing more and more energy along with an ever-growing population. This poses an imperative task for researchers to seek renewable clean energies as alternatives. In this context, many candidates fall into the searching scope, such as solar, wind, hydro, biomass, geothermal, hydrogen, and tidal energy sources. Among these possible sources for renewable energy, solar and wind are the most promising for providing electricity. Renewable electricity will be utilized in three primary domains: portable electronics, transportation, and power grids. However, solar and wind are by nature intermittent and geographically dispersed sources of energy [1, 2]. Consequently, effective utilization of these renewables will rely heavily on electrical energy storage (EES) devices.

Batteries, as EES devices, can store electricity in the form of chemical energy and are currently the only technically and economically viable EES devices for large-scale implementation of renewables [3]. Using batteries, surplus solar and wind power can be accumulated during their peak generation periods and then released whenever and wherever needed. Thus, batteries are critical to mitigate and ultimately solve the dual challenges of rising energy demands and global warming. Battery technology has evolved over several centuries, but since 1991 lithium-ion batteries (LIBs) have dominated consumer electronics, ascribed to their high energy and power. The first LIBs consisted of a graphite anode, a $LiCoO_2$ cathode, an electrically insulating separator, and an ionically conductive liquid electrolyte composed of a lithium salt dissolved in an organic solvent. Despite this dominance, state-of-the-art LIBs still lack the energy density and low cost necessary for use in transportation [1, 3]. For future transportation, energy densities 2 to 5 times greater are required to meet the performance goals of future plug-in hybrid-electric vehicles (HEVs) with a 40-80 mile all-electric range, and all-electric vehicles (EVs) with a 300-400 mile range [1]. Furthermore, existing battery technologies cannot meet the economic requirements for large stationary storage in power grids [2]. Thus, huge research effort is dedicated to pursuing next-generation LIBs and new battery technologies, such as lithium-sulfur (Li-S) [4, 5], lithium-oxygen ($Li-O_2$) [4, 6], and sodium-based batteries (NIBs) [7], with the goal of achieving high energy, high power, low cost, and reliable safety and stability. However, multiple complex technical issues must be solved for each of these battery systems. Quite often these issues arise when the electrode materials selected for high energy/power density (bulk properties), give rise to undesirable electrochemical interactions with the electrolyte (surface properties). A promising strategy to address these problems is by engineering the electrode surface using thin film coatings [8].

Among various coating technologies, atomic layer deposition (ALD) has recently emerged as a promising approach to achieve the aforementioned goals [9]. By applying a nanoscale surface coating on electrodes or separators, ALD can protect electrode materials from structural degradation and mitigate undesirable reactions at the interface between electrodes and electrolytes. Given its vapor-phase layer-by-layer nature, ALD is capable of producing uniform and conformal high-quality thin films compared to traditional sol-gel and other liquid-phase methods. In the past five years, ALD has received increasing attention, and many interesting studies have been conducted to improve battery performance. To date, ALD has greatly widened the applications of its nanoscale films from LIBs to Li-S, Li-O_2, and NIBs. At this stage, it is worthwhile to summarize the progress to date in order to stimulate further investigation into ALD batteries.

In the remainder of this Chapter, we illustrate the fundamentals and characteristics of ALD. As indicated by the title, the bulk of this Chapter will survey sub-nano to nanoscale ALD films for batteries and highlight various unique and novel ALD strategies for enhancing battery performance. In the last part, we offer outlooks for future research to inspire the continued development of ALD for high-performance batteries.

2. Atomic Layer Deposition and Thin Films

2.1. Surface Chemistry

ALD first emerged in the 1970s [10] and this invention is typically credited to Suntola and co-workers. Thereafter, ALD experienced relatively slow growth until the beginning of the 21st century when it was identified as a powerful vapor-phase tool for new nanotechnologies and novel nanomaterials [11]. Quite different from traditional source-controlled techniques such as chemical- and physical-vapor deposition (CVD and PVD), ALD is a surface-controlled process, and this enables accurate materials growth at the atomic level (typically at 1 Å growth per cycle) [12], low deposition temperatures (e.g., below 200°C) [13], fine-tunability over materials composition [14], and control over materials crystallinity [15]. These features also distinguish ALD from solution-based methods. In addition, ALD is simple in operation and efficient in materials usage. In terms of film quality, ALD enables sub-nano to nanoscale films with unrivalled uniformity and conformality [12]. The former is of particularly importance for producing pinhole-free films over large areas while the latter makes ALD an excellent technique for nanomaterials synthesis for a large range of application [12, 16-18]. Initially, ALD was developed for microelectronics [19, 20], but the range of applications has expanded greatly to include biomedical devices [21], catalysis [22], gas sensing [23], optoelectronics [24], nano and microelectromechanical systems (N/MEMS) [25], and energy technology [9, 26-30]. As a result of this proliferation of diverse applications, the list of ALD materials has grown enormously to include elements [15, 31-33], oxides [31, 32], sulfides [15, 30, 32], nitrides [15, 32, 33], fluorides [15, 32, 34], and complex compounds [15, 32]. To date, there are dozens of reviews published on the topics of ALD fundamentals and applications [12, 15-22, 24, 27-30, 32, 33, 35, 36].

Because ALD is surface-controlled, it is informative to consider an example of the surface chemical reactions responsible for ALD. In the case of Al_2O_3 ALD using trimethylaluminum (TMA) and water, the overall reaction is given by:

$$2Al(CH_3)_3 + 3H_2O \rightarrow Al_2O_3 + 6CH_4 \quad (1)$$

This overall reaction can be divided into two half-reactions as follows:

$$|\text{-}OH + Al(CH_3)_3\,(g) \rightarrow |\text{-}O\text{-}Al(CH_3)_2 + CH_4\,(g) \quad (2A)$$

$$|\text{-}O\text{-}Al(CH_3)_2 + 2H_2O\,(g) \rightarrow |\text{-}O\text{-}Al(OH)_2 + 2CH_4\,(g) \quad (2B)$$

where "|" designates species attached on the substrate surface and (*g*) signifies gaseous species. Each Al_2O_3 ALD cycle is composed of four steps: (i) TMA vapor is introduced and this reacts with surface hydroxyl (-OH) groups to form a new layer of $-Al(CH_3)_2$ and release CH_4 into the gas phase (Equation 2A); (ii) the TMA supply is terminated, and the system is purged to remove unreacted TMA and gaseous byproducts; (iii) H_2O vapor is introduced, and this reacts with the newly-formed $-Al(CH_3)_2$ species to produce Al_2O_3 and CH_4, and regenerate the hydroxyls (Equation 2B); (iv) the H_2O vapor is shut off and again the system is purged of residual H_2O and byproducts. This four-step ALD cycle can be repeated any number of times, and results in the layer-by-layer growth of Al_2O_3. The crucial aspect of ALD is that the two half reactions (2A and 2B) terminate automatically when all of the surface species (i.e., the –OH or $-Al(CH_3)_2$) are consumed by the ALD precursor. This characteristic ensures that each ALD cycle deposits a fixed amount of material on the surface, independent of precursor flux or substrate geometry, so long as the reactions are driven to completion. ALD precursors must be volatile and highly reactive with the co-reactant, but also chemically stable so as not to decompose in the gas phase or on the surface. The fact that the two precursors are supplied individually and react only on the surface allows the ALD reactions to be driven to completion, and this in turn provides lower deposition temperatures compared to CVD.

2.2. Characteristics of ALD

2.2.1. Atomic-scale Control over Stoichiometric Material Growth

As indicated by its name, ALD makes it possible to control material growth at the atomic level. A typical growth per cycle (GPC) is at the order of 1Å, and this value is dictated primarily by the surface density of functional groups and chemisorbed ALD precursor (e.g., the –OH or $-Al(CH_3)_2$ species). This differentiates ALD from other vapor-phase processes such as CVD and PVD. It needs to point out that the nature of precursors play important roles in determining material growth rate and the properties of the resultant materials. In the case of ALD Al_2O_3, many different precursors have been explored. Of them, the combination of TMA and H_2O is most widely used, and produces stoichiometric Al_2O_3 with a stable growth rate of ~ 1 Å/cycle over a relatively wide range of temperatures from room temperature (RT) to 200°C [13]. Using $AlCl_3$ and ROH (R = alkyl chains), however, the ALD growth rates are less than 0.7 Å/cycle and require higher temperatures (≥250°C). In some cases the ALD GPC can show a pronounced temperature-dependence. For instance, GaS_x ALD using $Ga_2(NMe_2)_6$ and H_2S shows a nearly decrease in GPC with temperature, and this behavior is attributed to the temperature-dependence of the concentration of surface functional groups [37]. In other cases, such as Li_2S ALD using lithium tert-butoxide (LTB, LiO^tBu) and H_2S where the concentration of functional groups is fairly independent of temperature, the GPC is found to be constant over a wide range (e.g., ~ 1 Å/cycle for Li_2S ALD in the range of 150 – 300°C). The strict requirement for surface functional groups to initiate the ALD process can sometimes lead to substrate-dependent growth. For instance, ALD can be substrate-inhibited or enhanced [32] in the nucleation regime at the start of growth, showing lower or higher GPC compared to the steady-state regime. These effects can be significant in battery applications utilizing sub-nm thickness films where the ALD is performed entirely in the nucleation regime. In cases where the substrate has an abundance of functional groups, and the concentration is essentially the same as in the steady-state regime for the ALD process, the GPC will be constant with ALD cycles.

2.2.2. Exceptional Uniformity and Conformality

One of the most attractive attributes of ALD is that it provides excellent thickness uniformity over large areas, such as the 300 mm diameter Si wafers used in microelectronics fabrication [38, 39]. Lupták et al. [38] reported that a 4 nm thick ALD HfO_2 film showed < 3% thickness variation on 300 mm Si wafers. To further distinguish itself from CVD and PVD, ALD is particularly skilled at depositing highly conformal films. In this case, conformality refers to the capability to produce films with uniform thickness inside of high aspect-ratio (AR) structures with minimal gradients. For instance, Elam et al. [40] demonstrated that ALD-ZnO conformally coated nanoporous anodic aluminum oxide of AR = 5000. In addition, ALD has been used for uniformly depositing conformal films on complex 3D substrates, such as graphene nanosheets (GNS) [41], forests of nanotubes/nanowires [42, 43], and others [44].

2.2.3. Low Temperature

Another major merit of ALD is that it permits low temperature deposition of less than 400°C in most cases. There are many studies reporting ALD processes at temperatures of less than 100°C (e.g., Al_2O_3 [13], TiO_2 [45, 46], ZnO [47], SnO_2 [48], etc.), and even room temperature (RT) (e.g., Al_2O_3 [13], B_2O_3 [49], and SiO_2 [50], etc.). This capability makes ALD suitable for thermally-sensitive substrates, such as polymers [51, 52] and biological templates [21, 53, 54].

2.2.4. Control of Materials Crystallinity

In spite of its low growth temperatures, ALD is particularly elegant in tuning material crystallinity either by adjusting temperatures, by selecting suitable precursors, or by using special substrates. It was reported that, using titanium isopropoxide (TTIP) and water, ALD produced amorphous TiO_2 at temperatures less than 150°C but crystalline anatase TiO_2 at 250°C [55, 56]. With an intermediate temperature of ~ 200°C, a mixture of both amorphous and anatase TiO_2 was received. Similar cases were observed with ALD processes for TiO_2 using $TiCl_4$ and water [57], and SnO_2 using $SnCl_4$ and water as precursors [41, 58]. In addition to temperature, precursor selection can influence the resultant material crystallinity.

Using TiI_4 and H_2O_2, for instance, the received TiO_2 could be tuned from anatase at below 300°C to rutile at above 325°C [59].

2.2.5. Wide Variety of Material Compositions

ALD is currently experiencing enormous expansion for new materials, enabling single elements, binary compounds (including oxides, nitrides, sulfides, phosphides, and fluorides), and even higher levels of complex compounds [15, 30, 31, 33]. In particular, the list is quickly growing with intensive investigations. As a consequence, ALD is providing new solutions to many realistic challenges, Now ALD has been widening its applications from the initial semiconductor industry [19, 20] to biomedical [21], catalysis [22], gas sensing [23], optoelectronics [24], N/MEMS [25], and energy technologies [9, 26-30].

3. Applications of ALD Surface Coatings in Advanced Batteries

3.1. Lithium-ion Batteries

3.1.1. Anodes

Lithium metal has a very high capacity of 3860 mAh/g, but the problem of dendrite growth prevents its use in commercial LIBs. The first LIBs were based on $LiCoO_2$ (LCO)/graphite, in which the former was used as the cathode while the latter used as the anode. With the desire for higher energy and power densities, new anodes and cathodes are under intensive investigation. Based on their electrochemical reactions with lithium ions, LIB anode materials can be classified into three categories: intercalation/de-intercalation (e.g., graphite), alloy/de-alloy (e.g., Si, Ge, and Sn), and conversion materials (e.g., metal oxides, sulfides, and fluorides). The materials in each class have their benefits and drawbacks. The typical LIBs of LCO/graphite are based on intercalation/de-intercalation. The chemical reaction on the graphite anode is as follows:

$$C + xLi + xe^- \leftrightarrow Li_xC_6\ (0 < x \leq 1) \tag{3}$$

in which lithium ions insert or extract during discharging and charging processes. Graphite as the most used anode features its flat and low working potential versus Li/Li^+, exceptional cycling life, abundance, and low cost. However, graphite has a low theoretical capacity of 372 mAh/g, owing to the fact that six carbon atoms accommodate only one lithium-ion at most. The diffusion rate of Li^+ into graphite is low, accounting for 10^{-9} to 10^{-7} cm^2/s [61]. In addition, the direct contact of graphite with organic electrolytes causes decomposition to form a layer of solid electrolyte interphase (SEI). In this context, many more materials as alternative anodes are under development, such as elements (e.g., Si, Ge, and Sn) [62], metal oxides (e.g., SiO, SnO_2, Fe_2O_3) [62, 63], metal sulfides [64], and metal fluorides [65]. Their main merit lies in much higher theoretical capacities over that of graphite, but they commonly suffer from high volume change, poor electron transport, capacity fading, and low Coulombic efficiency. In various efforts for addressing the aforementioned issues, surface coating has been remaining as an viable route [66], enabling reduction in side reactions, reliable electrochemical contact, and improved mechanical and electrical properties. Distinguished from traditional solution-based methods, ALD features itself a vapor-phase surface-controlled process and this enables ALD to modify electrodes via two technical routes: (i) forming an ultrathin film on individual powder particles of active materials and/or (ii) forming an integrated film on prefabricated electrode laminates. The two were verified being viable while the latter is much simpler. To differentiate them, in this Chapter name the two routes as "ALD-P" and "ALD-E", respectively.

3.1.1.1. Intercalation-Based Materials

Jung et al. [67] first reported ALD-Al_2O_3 for modifying graphite anodes via two routes: ALD-P and ALD-E. They found that a 5-cycle ALD-Al_2O_3 coating (~ 0.5 nm) via ALD-E could greatly boost the capacity retention of the coated graphite compared to the bare graphite anode (see Figure 1). In contrast, a 5-cycle ALD-Al_2O_3 coating via ALD-P degraded the graphite anode performance. They believe that the insulating layer of ALD-Al_2O_3 between individual graphite powder grains was responsible for the degradation. Furthermore, Jung et al. demonstrated that the benefit from a 5-cycle ALD-Al_2O_3 coating via ALD-E was even more significant at an elevated temperature of 50°C, accounting for > 90% retention of the modified anodes compared to 26% of the bare anode after 200 charge-discharge cycles. In a later study, Lee et al. [68] applied ALD-TiO_2 films via ALD-E on graphite anodes and found a layer of 400-cycle ALD-TiO_2 helped the modified anodes with better rate capability and cycling performance, due to its effect on suppressing charge-transfer resistance. A following work by Wang et al. [69] made a comparative study on ALD-Al_2O_3 and ALD-TiO_2 via ALD-E. Through measuring diffusion coefficients at different cyclic voltammetry (CV) scanning rates, they found that a 10-cycle ALD-TiO_2 coating could greatly improve the ionic diffusion kinetics of the coated graphite anodes, 4 times that of ALD-Al_2O_3 coated and bare graphite anodes (the two were comparable in ionic diffusion). They further demonstrated that a 3-nm ALD-TiO_2 on graphite anodes greatly improved the anode cycling stability and sustainable capacity at an elevated temperature of 55°C. Impedance spectroscopy showed that the ALD-TiO_2 coating stabilized the anode surface property and facilitated ionic transfer.

In addition to graphite, nanostructured carbon materials have been studied as alternatives, enabling larger surface areas and shorter Li^+ and electron pathways. However, their enhanced surface areas pose a big concern on SEI formation. Lahiri et al. [70] reported using ALD-Al_2O_3 to coat muti-walled carbon nanotubes (MWCNTs) grown by CVD on Cu foils. They demonstrated that the coated MWCNTs via ALD-E enabling exceptional rate capability and high capacity (3 times the theoretical capacity of graphite). In another study by Loebl et al. [71], the researchers used ALD-Al_2O_3 for coating carbon nanofibers (CNFs) via ALD-E and discovered that the CNFs as anodes showed an evident reduction in the first discharge capacity loss when they were coated by a 10-cycle ALD-Al_2O_3 layer. The study also disclosed that temperatures for ALD- Al_2O_3 films were not a key factor for the CNFs performance and the 10-cycle ALD-Al_2O_3 coated CNFs had capacities and rate capabilities comparable to the uncoated CNFs.

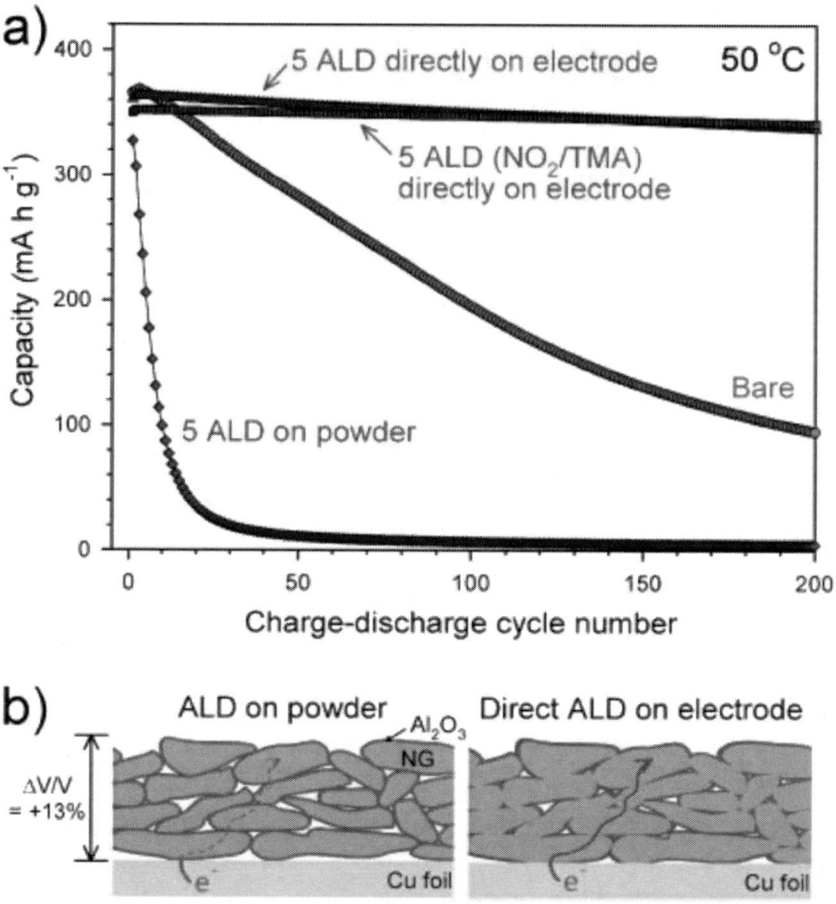

Figure 1. Electrochemical performance for ALD-coated NG composite electrodes: (a) cycle performance at 50°C and (b) schematic representation transport in NG composite electrodes prepared by ALD on powder and ALD directly on the electrode [67]. Reproduced by permission from Wiley.

Inspiringly, Piper et al. [72] recently investigated the effectiveness of ALD on carbon black (CB, a widely used conducting additive in LIB electrodes) via ALD-P. Their work demonstrated that a layer of 4 nm ALD-Al_2O_3 successfully inhibited the formation of SEI, greatly reduce the irreversible capacity loss, and improve Coulombic efficiency.

Besides carbon materials, $Li_4Ti_5O_{12}$ (LTO) also is an intercalation/de-intercalation anode material. In comparison to graphite, $Li_4Ti_5O_{12}$ has a spinel structure and is advantageous for its high working potential (~1.5 V vs. Li/Li^+). This high working potential makes $Li_4Ti_5O_{12}$ free of formation of SEI. Furthermore, this spinel is also more stable at elevated temperatures (75-80°C), for the electrode does not react with electrolyte components. In addition, $Li_4Ti_5O_{12}$ has minimal change in its lattice dimensions during Li^+ insertion/extraction, showing zero-strain during electrochemical cycling. The reversible intercalation/de-intercalation process is shown as follows:

$$Li_4Ti_5O_{12} + 3Li^+ + 3e^- \leftrightarrow Li_7Ti_5O_{12} \quad (4)$$

However, $Li_4Ti_5O_{12}$ suffers from a low electrical conductivity (5.8 × 10^{-8} S/cm at 140°C) [73]. Aimed at improving the interparticle electrical contact, Snyder et al. [73] applied 200-cycle (~5.8 nm) ALD-TiN on $Li_4Ti_5O_{12}$ NPs via ALD-P. TiN is a hard and refractory material having an excellent metallic-type conductivity. Electrochemical testing in the range of 1.4 – 2.2 V demonstrated that the coated $Li_4Ti_5O_{12}$ NPs showed dramatically improved stability, capacity, and rate capacity. It needs to particularly point out that this work represents the first ALD coating applied for battery studies.

In a following work, Ahn and Xiao [74] investigated the effect of a wider potential range of 0.001 – 2.5 V on the performance of LTO and demonstrated the effectiveness of ALD-Al_2O_3 coating via ALD-E for enhancing LTO performance. Extending the potential window to a voltage lower than 1 V may favorably offer a higher discharge capacity and a higher cell voltage for higher energy density of LIBs, but risk LTO with the capacity degradation due to the volume expansion and SEI formation. In the work, Ahn and Xiao demonstrated a continuous capacity fading with the bare LTO electrode, showing only half of its initial capacity after 100 cycles. In contrast, a 0.55 nm ALD-Al_2O_3 coating could significantly improve the LTO's cycling performance. In addition, the performance fading could be further mitigated with increased ALD-Al_2O_3 layers of 2.2 and 5.5 nm. The 2.2 nm coating helped harvest the highest sustainable capacity. On the other hand, the ALD-Al_2O_3 coatings reduced the initial capacity loss and increased Coulombic efficiency. Ahn and Xiao believed that the ALD-Al_2O_3 layers served as an artificial SEI to restrain the continuous reaction between electrolyte and LTO during charge-discharge cycling. Their EIS measurements demonstrated that the coatings helped sustain a stable interface between electrolyte and LTO, while the bare LTO was exposed to electrolyte to form a growing SEI which increased charge-transfer resistance. In another work using ALD-ZrO_2, Liu et al. [75] coated LTO nanosheets (NSs) synthesized by a microwave-assisted hydrothermal method. In the potential range of 0.1 – 2.5 V, the researchers revealed that a ZrO_2 coating of less than 5 cycles (~ 1 nm) could improve the sustainable capacity, cycling stability, and rate capability of the LTO NSs. Based on EIS measurements, they further verified that an ultrathin ZrO_2 coating (≤ 5 ALD cycles, 1 nm) could reduce charge transfer resistance occurred at the interface between electrolyte and LTO while suppress the formation SEI.

3.1.1.2. Alloy/de-Alloy Materials

Lithium alloying reactions in organic electrolytes were first reported by Dey [76]. Since then, metallic/semi-metallic elements and various compounds represent an important class of anode materials, including Si, Ge, and Sn, etc [77]. Among them, Si has been being pursued as one of the most attractive anode materials for LIBs, given its highest gravimetric (4200 mAh/g, $Li_{22}Si_5$) and volumetric capacity (9786 mAh/cm^3). In addition, Si is also abundant (the 2nd place in elements), cheap, and environmentally benign. The electrochemical reactions of Li with Si form intermetallic compounds such as $Li_{12}Si_7$, Li_7Si_3, $Li_{13}Si_4$, and $Li_{22}Si_5$ at distinct voltage plateaus, as described as follows:

$$Si + xLi^+ + xe^- \leftrightarrow Li_xSi,\ x \leq 4.4 \quad (5)$$

However, Si anodes are prone to experience serious capacity fading, owing to the huge volume expansion/contraction (~ 300%) during their charging and discharging processes. The big volume change can easily cause mechanical failures of electrode and ultimately cell termination [77]. In addition, electrolyte solvents are thermodynamically unstable below 1.5 V versus Li/Li+ and the decomposition products immobilize part of lithium in form of SEI. Due to the high volume change of Si, the SEI is prone to fracture during lithiation and de-lithiation, leading to exposure of fresh Si and increase of SEI with extended cycling. This increases the charge transfer and diffusion resistances.

Xiao et al. [78] and He et al. [79] almost at the same time first reported the effects of ALD-Al_2O_3 on Si anodes. Xiao et al. [78] found that a 5 nm ALD-Al_2O_3 coating could effectively suppress the formation of SEI, prevent electrolyte from decomposition, and improve Coulombic efficiency. In addition, they noticed that the insulating Al_2O_3 layer might have transferred into a more ionically conductive layer of $LiAlO_2$ after the first discharge. Furthermore, the ALD-Al_2O_3 layer greatly reduced the mechanical failure of Si nanofilms, showing much less micro-cracks. In the work, Xiao et al. analyzed the surfaces of both the uncoated and Al_2O_3-coated Si after 11 charge-discharge cycles using secondary ion mass spectrometry (SIMS). Their results disclosed that the SEI layer was 20-30 nm for the uncoated Si anode, mainly consisting of LiF, Li_2O, $LiCO_3$, and some P-containing species. In contrast, the ALD-Al_2O_3 coated Si anode only showed a SEI layer of a few nanometers, primarily compromising LiF and P-containing specie but little Li_2CO_3. They believe that the ALD Al_2O_3 layer could prevent electrons from reaching the surface to decompose the electrolyte, and this might have greatly reduced Li_2CO_3. X-ray photoelectron spectroscopy (XPS) analysis revealed that, after ~1 nm, two Al2p peaks appeared, one at 75.5 eV for Al_2O_3 and the other at 77.6 eV for AlF_3. Further deconvolution revealed another peak at 73.4 eV for $LiAlO_2$. They further disclosed that most of $LiAlO_2$ compound was found at the top surface and became weaker after 1 nm. $LiAlO_2$ is one type of solid state electrolyte. On the other hand, after 1 nm, the ratio between AlF_3 and Al_2O_3 was constant. The coexistence of AlF_3 of Al_2O_3 is believed to be beneficial for improving the ionic conductivity and the ion transference in the SEI layer. The AlF_3/Al_2O_3 formation is probably due to the reaction of PF_5 (resulting from the decomposition of $LiPF_6$) with trace water leading to HF which further reacts with Al_2O_3 and forms $AlF_3{\cdot}3H_2O$. In the work conducted by He et al. [79], they compared the performance of Si films and patterned Si columns, and found the latter could harvest much better cycling stability and much higher sustainable capacity. Applying a 2 nm ALD Al_2O_3 layer on the patterned Si columns, He et al. demonstrated improved capacity and Columbic efficiency, for the coating evidently reduced mechanical degradation of Si columns.

TiN has exceptional electronic conductivity and has little reactivity with Li [80]. Kohandehghan et al. [81] applied various thickness of ALD-TiN on Si nanowires (NWs) and compared the ALD-coated NWs with Si NWs coated by sputtered TiN. Their study demonstrated that a 5 nm ALD-TiN had the best effects on Si NWs in terms of capacity retention, Coulombic efficiency, and rate capability. They also disclosed that the ALD-TiN coating could greatly improve the Si NWs robustness mechanically and reduce the amount of SEI. The same group investigated another ALD coating of TiO_2 shortly on Si NWs [82]. Lotfabad et al. [82] applied both amorphous and crystalline TiO_2 on Si NWs using ALD, and they found that a 10 nm amorphous TiO_2 layer could best improve Si anodes' Coulombic efficiency (95% for bare Si NWs versus 99% for ALD-TiO_2 coated Si NWs), suppress SEI formation, and rate capacity. In addition, they also disclosed that TiO_2 partially changed into $LiTiO_2$ after one lithiation-de-lithiation. In a latest work by Lotfabad et al. [83], the researchers compared the effectiveness of TiO_2, TiN, and Al_2O_3 coatings on Si nanotubes

(NTs) and all the coatings were deposited by ALD. They found that it was most effective when Si NTs were coated both inside and outside. Of the three coatings, a 1.5 nm two-side ALD-TiO_2 coating could best enhance the electrochemical performance of Si NTs in Coulombic efficiency (up to 100%) and sustainable capacity (1700 mAh/g for two-side ALD-TiO_2 coated Si NTs versus 1287 mAh/g for bare Si NTs after 200 cycles at 0.2 C). It was verified that the TiO_2 coating consisted of amorphous TiO_2 and cubic $LiTiO_2$ after lithiation, which combine high ionic and electronic conductivity.

Through mixing Si nanoparticles (NPs) with polyacrylonitrile (PAN, 150000 g/mol) solutions, in a study by Li et al. [84] fabricated Si-PAN nanofibers (NFs). The received Si-PAN NFs were changed into Si-C (0.45:0.55 in ratio) NFs by carbonizing at a high temperature of 800°C. The researchers investigated the effects of ALD-Al_2O_3 on the electrochemical performance of the synthesized Si-C NFs. They found that ALD coatings could continuously improve the sustainable capacity and Coulombic efficiency of Si-C NFs with increased thickness when they were less than 28 ALD cycles. They believe that the ALD coatings offered benefits in both physical and chemical aspects. In the first place, ALD coating has strong physical/mechanical restraining effect on the Si-C composite NFs since the coating might transfer the stress of Si NP expansion from radial direction to in-plane restrain when the Si NPs are partly exposed on the surface. Secondly, ALD Al_2O_3 coating might act as a barrier for side reactions between the electrode and the electrolyte. The role of a chemical barrier combined with the reasonable mechanical properties makes the Al_2O_3 coating an artificial but strong and stable structure to improve the cycling performance and Coulombic efficiency of a Si-based anode. The important roles of ALD-Al_2O_3 were also verified in another study by Li et al. [85], in which Si-graphene (Si-G) composites via a CVD process were coated by 10-cycle ALD-Al_2O_3 in form of electrodes and thereby demonstrated improved reversible capacities. In particular, the ALD-Al_2O_3 coated Si-G composite anodes were demonstrated being effective in full cells with $Li_{1.2}Ni_{0.2}Mn_{0.6}O_2$ cathodes, showing very promising performance. In an effort for addressing the effects of surface coating on battery performance, Li et al. [86] utilized a modified potentiostatic titration technique (PITT) to quantify the charge-transfer kinetics of the ALD-Al_2O_3 coated Si electrodes. In their work, the electrochemical Biot number B was measured using the modified PITT, and can be viewed as the ratio of the solute (Li) diffusion resistance to that of the overall charge-transfer process. A large B number represents a system whose kinetics is limited by diffusion, whereas a small B number represents a system governed by the charge-transfer process. Their measurements revealed that, in comparison with the B number of the bare Si anode showing faster Li jump from the electrode to the electrode than the Li diffusion in the Si electrode, the B number increases with ALD-Al_2O_3 coatings up to 10 cycles (~1.1 nm). However, the B number decreases with thicker ALD-Al_2O_3 coatings. The researchers believed that the increasing B number for ALD-Al_2O_3 of less than 10 cycles is due to the formation of an artificial SEI and suppression of the electrolyte decomposition. When the artificial SEI is thicker than 1.1 nm, electrons transfer through this thick oxide layer would be much slower compared to the tunneling process through a thin oxide layer. As a consequence, for the case of thick coatings, the insulating nature of Al_2O_3 harms the kinetics and decreases the interfacial charge-transfer process, leading to decreased electrochemical B number.

Ye et al. [87] deposited ALD-Al_2O_3 and ALD-TiO_2 layers on Si micropillars and they clearly identified using SEM that the two coatings evidently mitigated the SEI formation. In contrast to the SEI layer of ~0.5-0.8 μm for bare Si pillars, the coated Si pillars showed the

SEI layer of < 150 nm. In addition, they also noticed that the bare Si circular-pillars showed relatively uniform volume expansion while the ALD-coated Si circular-pillars exhibited strong anisotropic expansion. In contrast, it was interesting that ALD-coated Si square-pillars became near circular shaper after lithiation.

Of various metal oxides, SnO_2 is a very promising anode material and has a high theoretical capacity of 781 mAh/g, over double that of graphite. The electrochemical reactions [77, 88] for SnO_2 are as follows:

$$SnO_2 + 4Li^+ + 4e^- \rightarrow Sn + 2Li_2O \quad (6)$$

$$Sn + 4.4Li^+ + 4e^- \leftrightarrow Li_{4.4}Sn \quad (7)$$

Apparently SnO_2 experiences two-step electrochemical process during the first discharge. SnO_2 first electrochemically reacts with Li^+ (see Equation 6) and the irreversible reaction produces Sn and Li_2O. The resultant Sn then further reacts with Li^+ (see Equation 7) to form the alloy $Li_{4.4}Sn$ and this step is electrochemically reversible. Due to the high ratio of Li to Sn in the reactions, there is a huge volume expansion (~300%) during alloying and a following large volume shrinkage during de-alloying.

Using a microwave-assisted hydrothermal method, Wang et al. [89] synthesized SnO_2 NPs of various sizes by adjusting post-annealing temperatures. The researchers made a comprehensive study on the effects of ALD-Al_2O_3 coating on the electrochemical performance of the SnO_2 NPs. They revealed that the optimal coating thickness via ALD-E is closely related to the SnO_2 particles sizes: the larger the particle size, the better the thicker ALD coating, vice versus. It is easy to understand that larger SnO_2 particles have larger volume change and therefore need a more robust coating to enhance their mechanical integrity and to remain their electrochemical contact. For example, for SnO_2 NPs of 4 nm, a 0.2 nm ALD-Al_2O_3 coating helped harvest the highest sustainable capacity and rate capability. In the case of SnO_2 NPs of ~25 nm, a 3 nm ALD-Al_2O_3 coating best improved the electrochemical performance.

In another work by Nguyen et al. [90], SnO_2 NWs were grown on stainless steel plates and then they were coated by ALD-Al_2O_3 and ALD-TiO_2 at 250°C. Using the synthesized and further coated SnO_2 NWs as anodes, the researchers investigated the effects of the coatings. They found that, in comparison a large irreversible capacity loss of the bare SnO_2 NWs in the first cycle, the SnO_2 NWs coated by either 1 nm ALD-Al_2O_3 or 5 nm ALD-TiO_2 did not show any capacity loss. This implies that the coatings could successfully suppress the formation of SEI. In addition, the coated SnO_2 NWs demonstrated greatly improved sustainability of high capacity and high rate capability. Besides acting as an artificial SEI, the coatings also served as a mechanical protective shield against pulverization during charge-discharge cycling.

In visualizing the effects of Al_2O_3, Liu et al. [91] utilized in situ transmission electron microscopy (TEM) to observe the pulverization of Al NWs. Al is an alloying/de-alloying anode material and experiencing ~100% volume expansion/contraction during discharge-charge processes. The Al NWs have a 4-5 nm thick Al_2O_3 layer. Liu et al. noticed that during the first lithiation Al turned into crystalline LiAl alloy with around 100% volume expansion. In the following first delithiation, nanovoids were formed and increased in size and number with further lithiation. This was due to the dealloying of lithium from the LiAl alloy. The researchers also noticed that the nanovoids could be partly healed in the second lithiation but the size and number of nanovoids increased in the following 2nd delithiation. After four charge-discharge cycles, there were nanovoids everywhere and the Al NW was turned into many isolated nanoparticles enclosed by a "glass tube". The researchers verified that the glass tube was larger than that of the original Al_2O_3 shell in diameter and also the tube wall thickness (7 nm) was larger than that of the original Al_2O_3 (5 nm) as well. In particular, the tube wall was stable after the first lithiation and did not shrink/expansion in the following cycles. The researchers confirmed that the tube wall was a Li-Al-O compound, serving as a solid electrolyte. The Li-Al-O glass exhibited good mechanical property, and did not show any crack or fracture. In addition, the glass layer provided a facile Li+ transport path relative to that in the usually formed SEI.

ZnO has a similar electrochemical mechanism as SnO_2, a conversion reaction followed by a alloying-dealloying reaction, described as below [92]:

$$ZnO + 2Li^+ + 2e^- \leftrightarrow Zn + Li_2O \tag{8}$$

$$Zn + Li^+ + e^- \leftrightarrow LiZn \tag{9}$$

ZnO has a theoretical capacity of 978 mAh/g, but suffers from a large volume change (~163%) and pulverization, leading to capacity fading during charge-discharge cycling. Aimed at reducing the reaction between electrolyte and ZnO nanorods (NRs), Lee et al. [93] applied a 3-10 nm thick ALD-TiO_2 polycrystalline anatase layer on the NRs, for TiO_2 could form a relatively stable SEI layer with electrolytes. They found that, in compared to the bare ZnO NRs, the coated NRs showed higher capacity and better cycleability. In another work, Yu et al. [94] synthesized ZnO-graphene nanocomposites via ALD in order to mitigate pulverization of ZnO and improve mechanical integrity of the resultant composites. Furthermore, Yu et al. [94] applied a layer of ALD-Al_2O_3 on the nanocomposites to suppress SEI formation and boost the electrochemical performance of the composite electrode. The researchers demonstrated that a 1 nm thick ALD-Al_2O_3 could reach the best electrochemical performance in terms of sustainable capacity and rate capability. After 100 charge-discharge cycling, they disclosed using scanning transmission electron microscopy (STEM) and energy-dispersive X-ray spectroscopy (EDS) that the uncoated electrode exhibited lots of large agglomerations of pulverized Zn/ZnO nanograins. In contrast, the 1 nm thick ALD-Al_2O_3 coated electrode still maintained the initial ultrafine particles of ZnO. Apparently, the ALD-Al_2O_3 coating protected ZnO nanoparticles from electrochemical pulverization and remained their morphology. Therefore, the coated electrode showed improved performance.

3.1.1.3. Conversion-Based Materials

Conversion-based materials represent an important class of battery anodes and cathodes and they are a new research scope in the 21st century [65]. The generalized reversible conversion reaction can be expressed as:

$$M_aX_b + (b\cdot n)Li \leftrightarrow aM + bLi_nX \tag{10}$$

Conversion-based materials could be transition metal oxides, sulfides, nitrides, fluorides, and even phosphides. For instance, Fe_3O_4 is a conversion-based material having a volume change of ~ 93%, and its electrochemical reaction can be described as follows:

$$Fe_3O_4 + xLi^+ + xe^- \leftrightarrow Li_xFe_3O_4 \quad (11)$$

$$Li_xFe_3O_4 + (8\text{-}x)Li^+ + (8\text{-}x)e^- \leftrightarrow 4Li_2O + 3Fe \quad (12)$$

Equation (11) is an intercalation/de-intercalation process while Equation (12) is a conversion reaction. Through confining Fe_3O_4 in porous carbon, Kang et al. [95] synthesized various Fe_3O_4-C nanocomposites in hope of improving electrical conductivity and harvesting better electrochemical performance. In comparison to micro and nanosized Fe_3O_4, the resultant Fe_3O_4-C nanocomposites showed improved cycling performance and higher sustainable capacity, but still suffered some capacity fading. Further applying a 0.2-0.4 nm thick Al_2O_3 ALD coating, the coated composites demonstrate much better electrochemical performance, due to enhanced mechanical integrity and reduced SEI formation.

3.1.2. Cathodes

In terms of structures, the primary cathodes in LIBs can be categorized into three classes: (i) layered $LiMO_2$ (M = Co, Mn, and Ni) as well as their derivatives; (ii) spinel $LiMn_2O_4$; and (iii) olivine $LiMPO_4$ (M = Fe, Mn, Co, Ni, Mg, or mixtures of them). Layered $LiCoO_2$ as a lithiated positive electrode material was first reported in 1980 [96]. It is also the first commercialized cathode in LIBs and features high capacity, good rate capability, and safety. However, if more than half of the lithium ions are extracted (Li_xCoO_2, $x < 0.5$; $V > 4.2$ vs. Li^+/Li), Co^{4+} will dissolve into the electrolyte, resulting in structural changes and thereby deterioration of the battery performance. The theoretical capacity of the $LiCoO_2$ is relatively low at ~ 130 mAh/g, for only around half Li/Co can be reversibly cycled without causing cell capacity loss due to changes in the $LiCoO_2$ structure [97]. Otherwise, two effects occur: phase changes that cause low reaction rates and the poor stability of the electrode at low lithium contents [97]. The high cost of Co, low capacity, and safety issues inhibit the application of $LiCoO_2$ in EVs and HEVs [97, 98]. Replacing a fraction of the Co with resource-abundant Ni and Mn could reduce the cost and improve structural instability. In contrast to $LiCoO_2$, however, much less research has been performed on $LiNiO_2$ and $LiMnO_2$. This is due to the issues related to their structural nature. $LiMnO_2$ is thermodynamically unstable and prone to convert into the thermodynamically stable spinel structure [97], while $LiNiO_2$ in the first place was not confirmed being existed stoichiometrically and unstable inherently, leading to potential dangers in contact with organic solvents [97]. An effective route for promising layered $LiMO_2$ developed is mixing different transition metals, and this has led to many more derivatives, such as $LiNi_{1/3}Co_{1/3}Mn_{1/3}O_2$, $LiNi_{0.5}Co_{0.2}Mn_{0.3}O_2$, $Li_{1.2}Mn_{0.55}Ni_{0.15}Co_{0.1}O_2$, and the likes. They could show higher capacity, more stable structure, or improved safety reliability.

In comparison to $LiCoO_2$, the spinel $LiMn_2O_4$ is one of the most promising cathodes because of its low material cost, high natural abundance, environmental harmlessness, and good safety [99]. However, $LiMn_2O_4$ suffers from fast capacity fading during cycling, mainly due to (i) dissolution of Mn^{2+}, (ii) Jahn-Teller distortion of Mn^{3+} ions due to the crystallographic transformation of $Li_xMn_2O_4$ ($1 < x < 2$) from cubic spinel structure to tetragonal phase that is well known for its poor capacity retention, and (iii) continuous decomposition of electrolytes on the electrode partially due to formation of Mn^{4+} that has high oxidation ability [100]. $LiFePO_4$ as the third generation of LIB cathodes, features its low cost, low toxicity, high safety, and good cycle stability, but is hindered by its low electronic and ionic conductivity.

In tackling the various issues associated with LIB cathodes, surface modification has proven to be simple and effective for improving physicochemical and electrochemical properties [8]. In 1999 Cho et al. [101-105] reported a pioneering series of papers showing that the capacity could be improved by coating a metal oxide or phosphate on the surface of $LiCoO_2$ particles. They found the capacity of the coated cathodes could be increased to 170 mAh/g when cycled between 2.75 and 4.4 V without capacity fade over 70 cycles. Since then, other researchers quickly confirmed the positive results of surface coatings. The mechanism of protection is related to minimizing the reactivity of Co^{4+} on charge with acidic HF in the electrolyte derived from the interaction of trace moisture with the electrolyte salt $LiPF_6$. Traditionally, solution-based methods such as sol-gel are widely used in earlier studies and the coatings on electrodes are from several tens or hundreds of nanometers to microns. The resultant coatings are not well controlled in thickness and uniformity. In addition, solution-based methods can only be applied to electrode material powders prior to fabricating into the laminated electrodes.

In the past several years, ALD was utilized as a new process to coat various cathodes and exhibited a series of benefits with demonstrable success. In the first place, ALD greatly saves coating materials and cost, showing comparable or better battery performance with sub-nano to nano ultrathin coatings. Secondly, ALD is the only one enabling to directly modify prefabricated electrodes via ALD-E with greatly improved performance. These well distinguish ALD from other methods and attract more and more research interest.

3.1.2.1. Layered Cathodes

$LiCoO_2$

Previously, surface modification via solution-based methods has demonstrated that metal oxides could help prevent Co dissolution and electrolyte decomposition [8]. Jung et al. [106] in 2010 delivered the first publication using ALD-Al_2O_3 to modify microsized $LiCoO_2$ (μ-$LiCoO_2$) and revealed that a layer of sub-nano (0.3 - 0.4 nm) Al_2O_3 could greatly boost the cathode cycleability and sustainable capacity. This has never been reported by other methods. Jung et al. [106] coated μ-$LiCoO_2$ powders using ALD Al_2O_3, and then the coated μ-$LiCoO_2$ were fabricated into electrodes. Electrochemical testing disclosed that a 2-cycle ALD-P Al_2O_3 coating (~ 0.3 – 0.4 nm) dramatically enhanced the cathode performance. After 120 charge-discharge cycles, the bare powders retained only 45% of their initial capacity while the 2-cycle ALD-Al_2O_3 remained 89% initial capacity. In the two cases, they had similar initial capacities. However, further increasing the ALD-Al_2O_3 coating thickness using 6 and 10 ALD cycles led to reduced sustainable capacity while the stability was highest with 10-cycle ALD Al_2O_3 (see Figure 2(a)). It was believed that the increased ALD-Al_2O_3 coatings should have resulted in difficulty for electron transport and slower Li^+ diffusion kinetics in the coatings. The two hindered the electrode to achieve high sustainable capacity with the 6 and 10-cycle ALD-Al_2O_3 coatings. In addressing the beneficial effects of the 2-cycle ALD Al_2O_3, the researchers conducted impedance measurements. The results revealed that the bared $LiCoO_2$ powders showed typical spectra comprising two semicircles and a 45° inclined line (see Figure 2(b)). The first semicircle in the higher frequency zone (indicated by #) is related to the solid electrolyte interphase (SEI) while the second semicircle in the lower frequency zone (signified by *) implies a charge-transfer reaction at the electrolyte/electrode interface. In comparison to the almost unchanged first semicircle, the second one of the bare $LiCoO_2$ powders increased dramatically in radius with electrochemical cycling, indicating an increase in charge-transfer resistance. In contrast, the single nonsymmetrical semicircle with the 2-cycle ALD-coated $LiCoO_2$ powders remained almost constant with electrochemical cycling, suggesting the interface between the electrode and the electrolyte was very stable and the charge-transfer resistance did not change. Jung et al. further postulated several mechanisms for the effectiveness of ALD Al_2O_3 [106]: (i) ALD Al_2O_3 helped suppress the structural instabilities related to lithium insertion and desertion; (ii) ALD Al_2O_3 may act as a solid electrolyte to prevent direct contact between the cathode and the electrolyte, implying ALD Al_2O_3 is less effective for electrolyte decomposition at higher potentials; and (iii) ALD Al_2O_3 may serve as a scavenger for HF resulting from the reaction of a trace H_2O with $LiPF_6$ in the electrolyte, reducing Co dissolution from $LiCoO_2$ due to the direct attack of HF.

Besides initiating the research on ALD coating on LIB cathodes, Jung et al. [106]for the first time proposed and verified that a direct ALD coating on prefabricated electrodes via ALD-E could achieve the similar effects as performed on cathode powders via ALD-P. Applying a 2-cycle ALD Al_2O_3 coating on a $LiCoO_2$ electrode, they demonstrated that this approach could achieve comparable battery performance. This method apparently is much simpler, and potentially more scalable for commercialization. In a shortly later work [107], the same group demonstrated a 2-cycle ALD Al_2O_3 coating on a nano-$LiCoO_2$ (n-$LiCoO_2$) cathode and realized a discharge capacity of 133 mAh/g under a high rate of 7.8 C (i.e., 1400 mA/g), accounting for a 250% improvement in reversible capacity compared to a bare n-$LiCoO_2$ electrode. Following Jung et al., other research teams [108-110] confirmed that nanoscale ALD Al_2O_3 films were viable for improving the performance of $LiCoO_2$ cathodes under different electrochemical conditions.

In addition to Al_2O_3, several other metal oxides also have been investigated, including ZnO [106], TiO_2 [109, 110], and ZrO_2 [110] in half cells. All of these studies demonstrated that Al_2O_3 performed much better in enhancing battery stability and capacity. Li et al. [110] showed that ALD coatings were more effective on the prefabricated $LiCoO_2$ cathodes via ALD-E than on $LiCoO_2$ powders via ALD-P. In a study of the underlying mechanism, Cheng et al. [109] revealed that ALD TiO_2 coatings were not stable during electrochemical cycling and participated in redox reactions. In contrast, ALD Al_2O_3 was stable. Based on their analysis, Cheng et al. [109] suggested that large band gap materials should be more promising and suitable as coatings of LIB cathodes.

Continuing their pioneered work, Jung et al. [111] recently investigated the effects of ALD Al_2O_3 in full cells, consisting of a μ-$LiCoO_2$ cathode and a natural graphite (NG) anode. In the work, they assembled bare/coated μ-$LiCoO_2$ cathodes and NG anodes into different full cells and found that a 2-cycle ALD Al_2O_3 coating was effective in any case (see Figure 3). Interestingly but surprisingly, they disclosed that a full cell consisting of a bare cathode (b-$LiCoO_2$) and a coated anode (c-NG) only showed a bit inferior to the c-$LiCoO_2$/c-NG cell. Using XPS and SIMS, they ultimately revealed that surface reactions occurring on the negative and positives electrode were related. In their work, they confirmed the presence of Al^+ nearby Co^+ on the cathode side in the cell of b-$LiCoO_2$/c-NG. They further explained that side reactions on the c-NG produced soluble Al-containing byproducts. These byproducts later diffused to the b-$LiCoO_2$ and contributed to forming a protective layer and reducing undesirable side reactions.

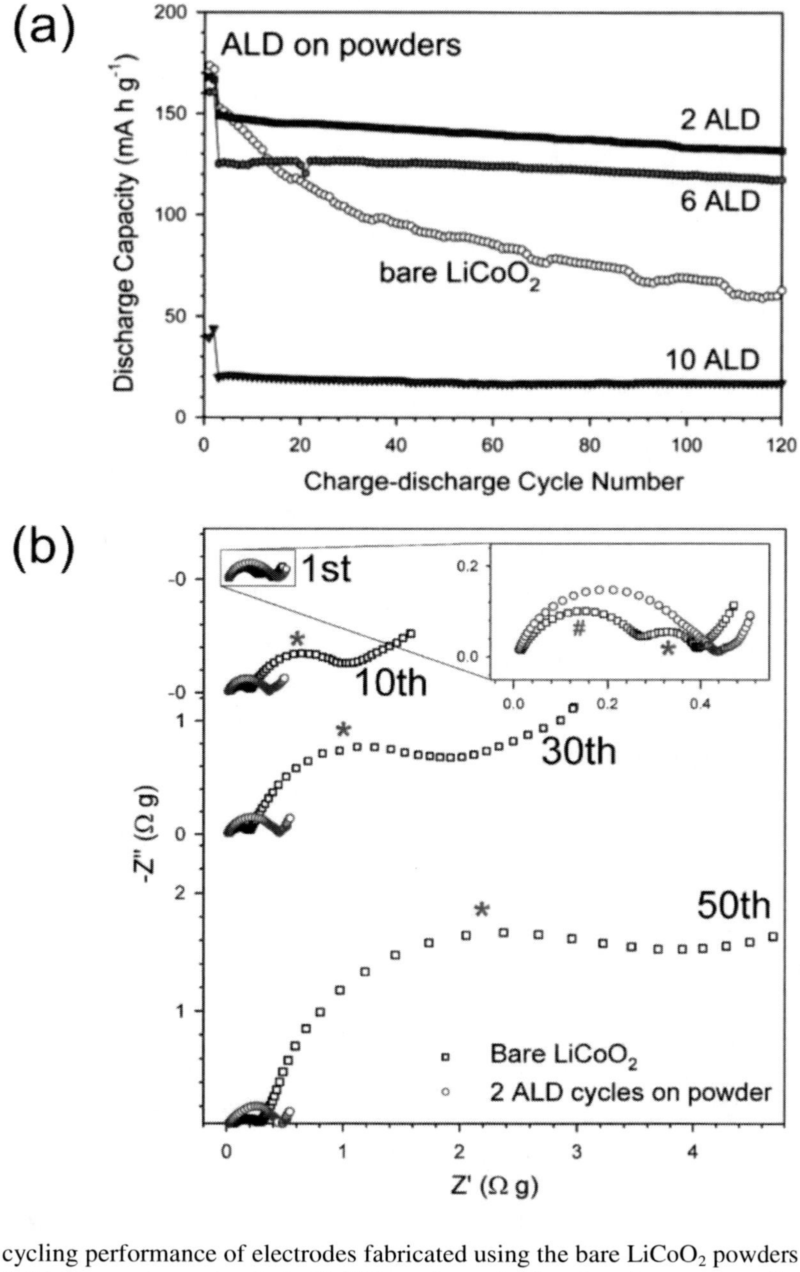

Figure 2. (a) cycling performance of electrodes fabricated using the bare $LiCoO_2$ powders and the ALD Al_2O_3 coated $LiCoO_2$ powders using 2, 6, 10 ALD cycles and (b) series of mass-normalized impedance spectra for electrodes fabricated with bare and ALD-Al_2O_3 coated $LiCoO_2$ powders for various charge-discharge cycle numbers [106]. Reproduced by permission from TheThe Electrochemical Society.

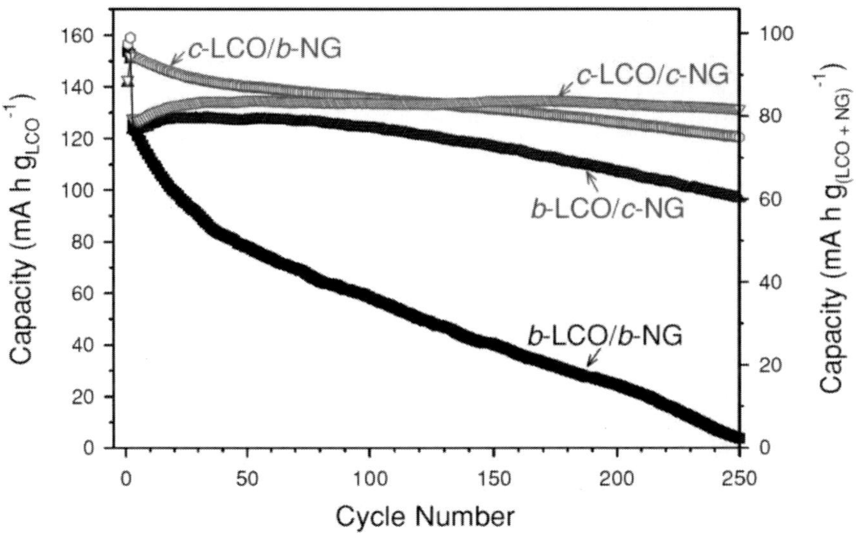

Figure 3. Cycling performance of $LiCoO_2$ / natural graphite (LCO/NG) full cells coated with ALD Al_2O_3. Reversible discharge capacity as a function of cycle number of four permutations of LCO/NG cells: Uncoated or bare electrodes are labeled with b, and ALD Al_2O_3 coated electrodes are labeled with c. Cycling tests were carried out between 3.25 – 4.45 V at 0.1 C for the first two cycles and 1C for subsequent cycles at room temperature [111]. Reproduced by permission from Wiley.

More recently, the team at University of Colorado studied full cells of solid-state lithium batteries using ALD Al_2O_3 [112]. In these cells, ball-milled $Li_{3.15}Ge_{0.15}P_{0.85}S_4$ and 77.5Li_2S-22.5P_2S_5 (mol%) were used as solid state electrolytes. The former was mixed with carbon additives, and bare/ALD Al_2O_3-coated $LiCoO_2$ to produce composited cathodes. The coated $LiCoO_2$ powder were first heated under 300°C for 12 hours in the presence of Ar. Then the composite cathodes were then assembled with another solid state electrolyte of 77.5Li_2S-22.5P_2S_5 and Li metal into full cells. The work demonstrated that the 2-cycle and 4-cycle ALD Al_2O_3 coated $LiCoO_2$ composite cathodes were comparable in performance and much better than the uncoated one in cycling stability and sustainable capacity. A 6-cycle ALD Al_2O_3 coated $LiCoO_2$ composite cathode showed a worse performance. The authors confirmed that a heating treatment was essential for the coated cathodes to achieve their better performance, leading to better Li^+ intercalation to $LiCoO_2$.

Very different from the metal oxide coatings discussed above, our team developed a new coating material, $AlWF_x$, using WF_6 and TMA as the ALD precursors [113]. This new coating material had exceptional electrical conductivity (3.5×10^{-2} Ω·cm). Applying this metal fluoride composite on $LiCoO_2$ cathodes directly, Park et al. [113] demonstrated that the 1 nm coated cathode showed excellent rate capability. At a current density of 400 mA/g, the discharge capacities were 51% and 92% of the first capacities for bare and coated cathodes, respectively. The two cathodes had no difference in their first initial capacities.

$LiNi_{1/3}Co_{1/3}Mn_{1/3}O_2$

$LiNi_{1/3}Co_{1/3}Mn_{1/3}O_2$ (NCM) cathode material features a high capacity and improved electrical conductivity. This mixed transitional metal ion cathode effectively integrates the benefits of $LiCoO_2$, $LiNiO_2$, and $LiMnO_2$. Thus, it has been undergoing extensive studies. Earlier studies revealed that NCM suffers from drastic degradation in capacity and cycling life at elevated temperatures (45 -60°C) and voltages (> 4.3 V). By applying a 4-cycle ALD Al_2O_3 on NCM powders and then annealing for 12 hours at 300°C in air, Kim et al. [114] demonstrated that the coated NCM powders were similar to the bare ones in capacity and sustainability in 100 charge-discharge cycles at a rate of 1 C (i.e., 160 mA/g) at room temperature. However, they showed a much higher capacity of 118 mAh/g compared to 35 mAh/g for the bare ones at a higher rate of 4 C. The outstanding rate capability of the coated NCM was very encouraging and confirmed under different rates up to 16 C, where ~40 and 0 mAh/g for the coated and bare ones, respectively. Kim et al. ascribed the exceptional performance of the coated NCM to the modification of the ALD Al_2O_3 on the surface characteristics of the NCM powders, and they confirmed that both the resistance of the surface films at the electrode/electrolyte interface and the charge-transfer resistance were stable and small for the coated NCM, while the two increased drastically with increased charge-discharge cycling. Using TEM, Kim et al. [114] revealed that, after 100 charge-discharge cycles, the bare NCM powder underwent phase transition from layered structure mother phase to spinel cubic structure. In contrast, the coated NCM powders preserved their original layer structure. In other words, the excellent rate capability of the coated NCM was benefited from their sustained structure. This merit of the ALD Al_2O_3 was further examined at an elevated temperature of 55°C. Results showed that, under 1 C, the coated NCM could remain 92% of its initial capacity, 156 mAh/g, while the bare NCM could only hold 14% of its initial capacity, 24 mAh/g.

In another study, Li et al. [115] applied a new coating material of $LiTaO_3$ directly on pre-fabricated NCM electrodes with different thickness from 2, to 5, 10, 20 ALD cycles. The $LiTaO_3$ is a solid state electrolyte having better ionic conductivity than ALD Al_2O_3. Li et al. demonstrated the $LiTaO_3$ coatings could greatly improve the electrochemical performance of the NCM cathodes in sustainable capacity and stability. Of all the thickness of the ALD coatings, 5 cycles were the best in most of the cases including increased cut-off voltages from 4.3 – 4.8 V, high rates, and an elevated temperature of 55°C. The coating layers were confirmed being able to provide stable interfaces between the NMC/electrolyte, while the interfaces of the bare NCM electrodes evolved with charge-discharge cycling.

$LiNi_{0.5}Co_{0.2}Mn_{0.3}O_2$

Layered $LiNi_{0.5}Co_{0.2}Mn_{0.3}O_2$ (NCM-523) is one representative kind of ternary material $LiNi_xCo_yMn_{(1-x-y)}O_2$, featuring its nickel-rich characteristics. This cathode has a capacity of ~200 mAh/g. However, it suffer from interfacial stability issues between the electrode and electrolyte. Kong et al. [116] applied 8-cycle ALD ZnO on NCM-523 powders and the coated NCM-523 demonstrated dramatically improved capacity and rate capability. EIS measurements revealed that the coated NCM-523 showed very stable interface indicated by almost unchanged impedance while the bare NCM-523 suffered from increasing impedance with extended cycling.

$Li_{1.2}Mn_{0.55}Ni_{0.15}Co_{0.1}O_2$

Among the advanced cathode materials, lithium- and manganese-rich (LMR) oxides, $xLi_2MnO_3{\cdot}(1-x)LiMO_2$ (M = Mn, Ni, and Co), are promising in terms of safety and energy density. However, LMR composites must be charged to above 4.5 V to fully activate the Li_2MnO_3 component. The activated LMR can reach 250 mAh/g at C/10. In addition, the wide operating voltage (2.0 – 4.7 V) poses issues for long-term stability due to the decomposition and dissolution of transition metals occurred at the electrode/electrolyte interface. In addition, LMR cathode may form a spinel-like phase during cycling, which can lead to uncontrolled voltage changes of the LMR material during their life span.

Jung et al. [117] first reported using ALD Al_2O_3 to modify $Li_{1.2}Mn_{0.54}Ni_{0.13}Co_{0.13}O_2$ (LMNCO) laminates directly. They disclosed that the coated electrodes could be improved in performance using several cycles of Al_2O_3. They further found that a heat treatment at 300°C for 3 hrs could further enhance the coated cathode performance, for the treated coating showed better Li-ion kinetics.

In a recent work by Zhao et al. [118], the researchers synthesized LMNCO nanoparticles. By coating the nanoparticles with ALD Al_2O_3, ZnO, and ZrO_2 via ALD-P, the team successfully improved the LMNCO performance in capacity and sustainability. Of the three ALD coatings, ZrO_2 outperformed ZnO and Al_2O_3. Realized the low Li^+ conductivity of ZrO_2, the researcher developed a core-shell structure of LCO@LMNCO in which LMNCO was covered by a LCO shell, and the 10 nm LCO provided the best performance. Further coating the core-shell LCO-10@LMNCO nanoparticles with a 1 nm ALD ZrO_2 film, the resultant cathode received a much higher capacity with exceptional sustainability, from the initial discharge capacity of 259.8 mAh/g to the 184.0 mAh/g at the 100th cycle. In contrast, the bare LMNCO changed from the 1st 114.5 mAh/g to the 100th 65.9 mA/g.

Zhang et al. [119] deposited both ALD 2-3 nm Al_2O_3 and TiO_2 thin films on $Li_{1.2}Mn_{0.55}Ni_{0.15}Co_{0.1}O_2$ powders to tune the interface between the LMNCO cathode and electrolyte. In electrochemical testing on half cells using Li metal as anodes, they disclosed that the ALD Al_2O_3 could boost the cathode capacity and stability at room temperature and an elevated temperature of 55°C. In comparison, ALD TiO_2 could not help improve the cathode performance, due to the redox of TiO_2 in the range of 2-4.8 V. In their following testing on full cells (using MCMBs as anodes), however, they found that there were no evident differences between the coated and bare cathode.

In another study by Bettge et al. [120], they applied ALD Al_2O_3 on prefabricated LMNCO cathodes via ALD-E. Results from electrochemical testing on full cells demonstrated that a 3.4 nm thick ALD Al_2O_3 could harvest the best performance in terms of capacity and sustainability. They also found that the coating could reduce electrolyte decomposition and thereby stabilize the cell impedance. In addition, the coating could inhibit the dissolution of Mn, Ni, and Co, but not entirely prevent. In a recently comprehensive comparative study by Bloom et al. [121], a LMR-NMC was coated by various coatings including Al_2O_3, ZrO_2, TiO_2, and ALD $LiAlO_x$ via ALD-E. The results showed that ALD $LiAlO_x$ could help get the highest capacity with the best sustainability. However, all the coatings tested could not help sustain the cathode structure. Zhang et al. [122] developed a thin ALD Al_2O_3 film on another Li-rich cathode, $Li_{1.2}Ni_{0.2}Mn_{0.6}O_2$ and they found that the coating could not help avoid the cathode voltage fading.

3.1.2.2. Spinel

$LiMn_2O_4$

Guan et al. [123] first reported ALD Al_2O_3 for surface-modifying the spinel $LiMn_2O_4$ (LMO), and disclosed that a 2 nm coating is very beneficial for sustaining the cathode capacity and stability using Ag as the reference electrode. In one of their following studies

[124, 125], Luan et al. [124] verified that ALD could realize uniform coatings on both microsized and nanosized LMO powders via ALD-P. Luan et al. [124] demonstrated that the n-LMO performed much better than the μ-LMO, owing to their larger surface area and smaller size at an elevated temperature of 55°C. This is particularly evident under a high rate of 5 C and a 0.72 nm ALD Al_2O_3 coating could greatly enhance the sustainable capability. Their investigation showed that a coating on prefabricated electrodes via ALD-E was more effective than one on cathode powders via ALD-P. This might be due to several facts: (i) the ALD Al_2O_3 served as HF scavenger and protected LMO from HF attack; (ii) the ALD Al_2O_3 layer might acted as a solid state electrolyte and hindered the liquid electrolyte from decomposition; and (iii) the ALD coating on electrodes provided a solid framework, enhanced mechanical robustness, restrained the phase transformation of LMO from cubic to tetragonal structure, suppressed harmful Jahn-Teller effects, and improved electrochemical contact. In comparison, the ALD coating on powders might have increased charge transfer resistance and slowed down electron transport between LMO particles and carbon additives.

In addition to Al_2O_3, ZrO_2 [126-128] and ZnO [128, 129] were also reported for modifying LMO. Different for the ALD Al_2O_3 effects, both the ALD ZrO_2 and ZnO coatings could perform the best when they were coated on LMO powders via ALD-P but not on LMO electrodes via ALD-E with a thickness of ~ 1 nm. In addition, it was demonstrated that a post-annealing (450°C for 3 hours in air) on ALD ZrO_2-coated LMO powders could improve the cathode performance in sustaining high capacity and stability at both the room temperature of 25°C and an elevated temperature of 55°C [127]. In a comparative study [128], Zhao and Wang demonstrated that ALD ZnO had the best enhancement on LMO performance, followed by ZrO_2 and Al_2O_3. Besides the similar effects as ALD Al_2O_3 as discussed above, the excellent electronic conductivity of ZnO and ZrO_2 might have made them better than the insulating Al_2O_3 when they were coated on LMO powders.

$LiNi_{0.5}Mn_{1.5}O_4$

$LiNi_{0.5}Mn_{1.5}O_4$ (for short, LNMO) is a high-voltage cathode material (4.7 V vs. Li/Li^+) and a derivative of the commercialized spinel LMO, aimed at suppressing the intrinsic deficiencies such as Jahn-Teller distortion of Mn^{3+} [98] with the same theoretical capacity as LMO (147 mAh/g). Fang et al. [130] for the first time applied ALD Al_2O_3 to directly modify LNMO electrodes and they demonstrated a layer of 1 nm ALD Al_2O_3 coating could greatly boosted the performance of the modified LNMO cathode via ALD-E at an elevated temperature of 55°C. After 100 cycles at C/2 (1C = 140 mA/g), the ALD Al_2O_3 coated LNMO delivered 116 mAh/g, corresponding to 90% of the original capacity, but the bare LNMO could only maintain 98 mAh/g, which is less than 83% of the initial capacity. After 900 cycles, the ALD Al_2O_3 coated LNMO showed 63% capacity retention, accounting for only 0.04% capacity decay per cycle. The excellent performance of the modified LNMO was ascribed to the beneficial effects of the ultrathin 1 nm ALD Al_2O_3 coating.

This was confirmed in another following work by Kim et al. [131]. Instead of coating electrodes, Kim et al. deposited ALD Al_2O_3 on LNMO powders with 2, 4, 6, and 10 cycles. Compared to bare LNMO powders, the coated LNMO powders via ALD-P did not help improve the cathode capacity but greatly boosted Coulombic efficiency. In addition, impedance measurements revealed that the coated LNMO had much stable resistance while the bare LNMO continued to increase with cycling. In addition, they demonstrated that the 4-cycle ALD Al_2O_3 coated LNMO via ALD-P showed less self-discharge in comparison to the bare LNMO. These demonstrated that the ALD Al_2O_3 coating could significantly suppressed the undesirable side reaction on the LNMO surfaces. *Ex situ* XPS analyzed both the bare and 4-cycle ALD Al_2O_3 coated cathode after 50 charge-discharge cycles and the results confirmed that there were much thinner SEI formed on the coated cathode. Due to the attack of HF, the ALD Al_2O_3 layer changed into a matter of AlO_xF_y. In addition, *ex situ* SIMS depth profiling verified that there were much less organic component on the Al_2O_3 coated LNMO than the bare LNMO. In particular, fuel cells of LNMO/$Li_4Ti_5O_{12}$ clearly demonstrated that the full cell with the ALD Al_2O_3 coated LNMO outperformed the one with the bare LNMO. This was believed that the ALD Al_2O_3 coating enhanced the stability and the higher Coulombic efficiency.

Xiao et al. [132] conducted a study on LNMO/graphite full cells in which graphite anodes or LNMO was coated via ALD-E with thin Al_2O_3 coatings. They revealed that, compared to a relatively stable performance of LNMO in half cells, LNMO/graphite full cells exhibited serious performance fading. Solely applying ALD-Al_2O_3 on LNMO cathodes, the resultant full cells demonstrated improved capacity. Surprisingly, solely applying ALD-Al_2O_3 on graphite anodes, the resultant full cells could even show much higher sustainable capacity. XPS analysis and TEM observation on graphite anodes jointly disclosed there was the least amount of Mn from the cell with the solely coated graphite anode, followed by the cell with the solely coated LNMO cathode and the cell with uncoated LNMO and graphite electrode. Xiao et al. [132] speculated a possible explanation: the Al_2O_3 coating might have passivated some active sites (such as –OH and –COOH functional groups which were consumed during ALD-Al_2O_3 coating) on the graphite and carbon black surfaces, leading to higher energy barrier. This might have made nucleation of metallic Mn nanoparticles more difficult. Also, reduction of Mn requires electron to be transferred to the top surface of graphite to meet Mn ions. The ALD-Al_2O_3 on the graphite is an insulating layer, which blocks the electron transfer from the current collector to the top surface of the graphite to reduce Mn^{2+}. However, as Xiao et al. pointed out, the exact mechanism is still unclear. Xiao et al. also noted that the SEI on the Al_2O_3-coated graphite particle surfaces is much thinner than that on the uncoated graphite. This indicated that the Al_2O_3 coating also suppressed the electrolyte decomposition, leading to less SEI formed and better capacity retention. In another research attempt [133], recently we made a comparative study on full cells consisting of LNMO and graphite, in which Al_2O_3 and $LiAlO_x$ were used to modify the electrodes via ALD-E. The $LiAlO_x$ was verified having much higher ionic conductivity than Al_2O_3, and is a solid state inorganic electrolyte. Electrochemical testing on various full cells disclosed that either $LiAlO_x$ or Al_2O_3 was effective for greatly improving the cell performance. In all cases, the full cell consisting a 2-cycle ALD $LiAlO_x$ coated cathode and an uncoated graphite anode demonstrated the best performance. XAS analysis detected both Mn^{2+} and Ni^{2+} on the graphite anode in all the cases, but a 0.2 nm $LiAlO_x$ or Al_2O_3 coating on the graphite anode could greatly inhibited the dissolution of both Mn^{2+} and Ni^{2+}.

Besides the oxides discussed above, a recent study [134] applied ALD amorphous $FePO_4$ to modify LNMO powders. Xiao et al. revealed that the 1 nm ALD $FePO_4$ coated LNMO powders exhibited the best performance in terms of capacity, sustainability, and rate capability. The researchers confirmed that the ALD $FePO_4$ coating could inhibit the side reactions on the LNMO cathode by avoiding the direct contact of LNMO with the electrolyte. In addition, the ALD $FePO_4$ coating could help accommodate lithium ions rapidly during cycling.

3.2. Beyond Lithium-Ion

3.2.1. Lithium-Sulfur Batteries

Li-S batteries are currently attracting great research interest, owing to the high capacity (1672 mAh/g), abundance, cost-effectiveness, and environmentally benignity of S. Li-S batteries theoretically offer 4-5 times higher energy density over LIBs, and are sufficient to power EVs. However, Li-S batteries suffer from several serious technical issues [4, 5]: (i) electronically and ionically insulating nature of S and Li_2S; (ii) shuttling of polysulfides (Li_2S_n, $4 \leq n \leq 8$, intermediate products during charge-discharge cycling) from the cathode side to the lithium metal anode side, resulting in undesirable side reactions on Li metal; and (3) safety risk posed the use of lithium metal as anode.

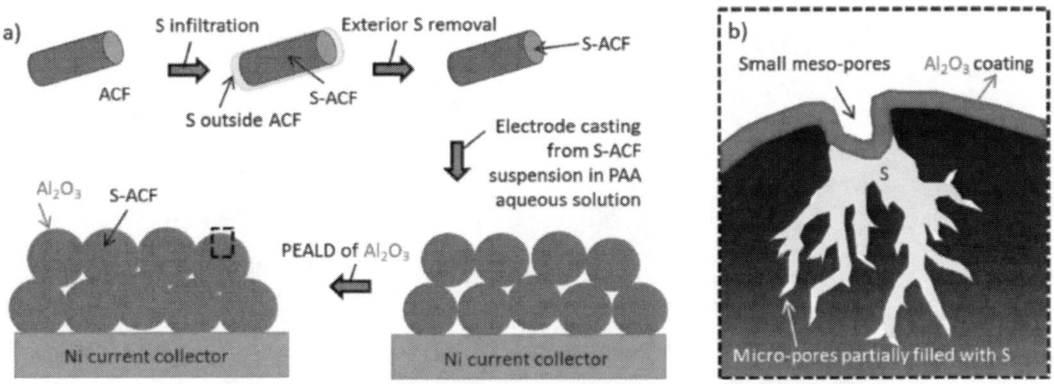

Figure 4. Schematic of PEALD-coated S-ACFs electrode preparation: (a) steps involved in the preparation process, and (b) high-magnification schematic showing conformal alumina coating around S-ACF particles exposed to gaseous environment during the PEALD deposition [135]. Reproduced by permission from Wiley.

Using a plasma-enhanced ALD (PEALD) for Al_2O_3 of TMA and O_2 at 170°C, Kim et al. [135] coated S-infiltrated activated carbon fibers (S-ACFs) via ALD-E (see Figure 4), in which S took 50% by weight. They demonstrated that a 5 nm Al_2O_3 layer greatly improved the electrode performance by forming a physical barrier to mitigate S shuttling and thereby to reduce the side reactions at an elevated temperature of 70°C, accounting for a high specific capacity of 600 mAh/g after 300 charge-discharge cycles at 0.2 C. Similarly, Yu et al. [136] used ALD-Al_2O_3 of TMA and water for coating a composite of S-graphene oxide (GO) via ALD-E at 120°C, in which S was around 50% by weight. The researchers revealed that a 10-cycle ALD-Al_2O_3 boosted the retention of high capacity of the coated electrode, accounting for 82% compared to 45% of the bare composite S-GO electrode. They believed that the ALD coating formed a layer of physical barrier which inhibited the diffusion of polysulfides.

3.2.2. Lithium-Air Batteries

Li-O_2 batteries are another type of high energy EES devices enabling even greater energy density (3623 Wh/kg when discharged to Li_2O_2 and 5204 Wh/kg when discharged to Li_2O) than Li-S batteries, but facing many technical challenges [6]. Among them, one big concern is to lower their high overpotential during charge. One route for tackling this issue is using nanosized catalysts. Using ALD, Lu et al. [137] and Lei et al. [138] developed a new conceptual nanostructured cathode for Li-O_2 batteries in which ALD-Al_2O_3 was used to protect electrolytes from decomposition and ALD-Pd applied to deposit Pd nanoparticles as electrocatalysts for promoting the formation of nanocrystalline Li_2O_2 (see Figure 5). In the study, super P conductive carbon black was used the catalyst support. The carbon support was first deposited with a layer of 3-cycle Al_2O_3 via ALD-P to passivate the defects on the carbon surface and then a 3-cycle ALD-Pd was further deposited to form Pd nanoparticles as catalysts. The coated carbon black then was demonstrated significantly lowering the charge potential from 4.4 to 3.2 V, while changing the discharge potential from ~2.5 to ~2.7 V. As a consequence, the overpotential of the Li-O_2 batteries was greatly reduced, attributing to the protective layer of ALD-Al_2O_3 and the electrocatalysts by ALD-Pd.

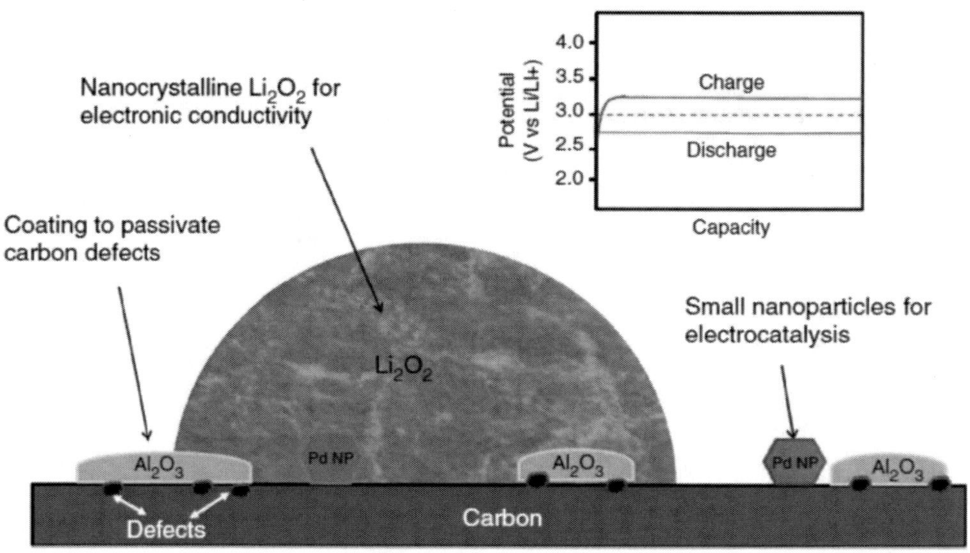

Figure 5. Schematic of the nanostructured cathode architecture showing the ALD-Al_2O_3 coating, the ALD-Pd nanoparticles, and the nanocrystalline Li_2O_2, all of which contribute to lowering the overpotential of Li-O_2 batteries [137]. Reproduced by permission from Wiley.

3.2.3. Sodium-Ion Batteries

Besides the new trends in Li-S and Li-O_2 batteries, there was some study on ALD for NIBs recently. NIBs are similar to LIBs but replace Li anodes by Na anodes, emphasizing the abundance of Na resources [7]. In a latest work, Han et al. [139] developed an electrode by depositing Sn NPs on carbon fibers (CFs), and they further applied a thin layer of 60-cycle ALD-Al_2O_3 on the resultant Sn-CFs composite electrode. Sn suffers from serious volume change during charge-discharge cycling, leading to fast capacity fading. Han et al. revealed that the ALD-coated Sn-CFs could remain exceptional capacity retention, accounting for a stable capacity of 650 mA/g in comparison to 110 mA/g for the bare electrode at the 40th cycle. Using in situ TEM, Han et al. clearly demonstrated that the ALD coating greatly improved the mechanical property of the coated Sn-CFs and helped sustain excellent electrochemical contact during charge-discharge cycling. In addition, they showed that the initial Al_2O_3 coating changed into Na-Al-O after the first sodiation. This Na-conductive layer was not reversible but the core of Sn NPs were reversible during charge-discharge cycling. Thus, the ALD-Al_2O_3 played multiple roles in the NIBs.

CONCLUSION AND OUTLOOK

Surface coating of battery electrodes has emerged as a promising strategy for tailoring electrode-electrolyte interfaces in order to improve their chemical stability and thereby to increase battery performance. ALD as a vapor-phase thin film technique recently demonstrated tremendous abilities over traditional methods for greatly enhancing battery performance, featuring its capabilities enabling a large variety of sub-nano to nanoscale uniform and conformal coatings. In addition, ALD is to date the only viable technique that allows thin films to be deposited directly on electrodes. These distinguish ALD as an exceptional route to improve battery safety, cycling stability, and capacity. Besides their widely demonstrable successes in LIBs, ALD is now being investigated as an important approach in various new battery technologies for achieving desirable performance, including Li-S, Li-O_2 batteries, and Na-ion batteries.

The roles of ALD ultrathin coatings vary in different battery systems. In LIBs, ALD coatings are responsible for: (i) reducing side reactions as artificial SEI at the interfaces between an electrode (an anode or cathode) and electrolyte; (ii) stabilizing electrode structures; (iii) serving as a scavenger for HF resulting from the reaction of a trace amount of H_2O and $LiPF_6$ in the electrolyte; (iv) stabilizing the chemical composition of electrolytes; (v) enhancing mechanical robustness of electrodes; and (vi) sustaining excellent electrochemical contact of electrode active materials.

In Li-S batteries, the main role played by ALD-Al_2O_3 layers was formation of a physical barrier for mitigating the diffusion of polysulfides and then reducing undesirable reactions on the lithium metal anode. In Li-O_2 batteries, ALD coatings could be artificial SEI to protect electrolytes from decomposition or catalysts to reduce overpotentials. In the latest study for NIBs, ALD Al_2O_3 was demonstrated to be beneficial for reducing SEI formation, sustaining excellent electrochemical contact, enhancing mechanical robustness, and providing a Na-conductive interface.

ALD is a versatile technique for providing multiple solutions to technical challenges in various battery systems. Its features lie in its atomic-scale accuracy, high-quality films, viability for electrodes directly, and function-oriented coating selections. However, it needs to particularly point out that the mystery of ALD coatings is still remaining and its exact mechanisms are still unclear. In addition, we need to be aware that there still is a large space unexplored for ALD coatings to fully optimize various battery systems and future research efforts are urgently needed. To date, only limited ALD coatings have been investigated and we may achieve much better battery performance in future work. Furthermore, the scale-up of the ALD technique for battery studies is remaining questionable and this is very critical for ALD being used as a realistically useful method for developing next-generation advanced EES devices. Despite of these, the studies in literature have proven ALD being promising and unique in battery research. Thus, ALD holds great promise for helping accomplish our goals in EVs and large-scale stationary storage devices.

ACKNOWLEDGEMENTS

This work was supported as part of the Joint Center for Energy Storage Research (JCESR), an Energy Innovation Hub funded by the U.S. Department of Energy, Office of Science, Basic Energy Sciences (BES). Electron microscopy was performed at the Electron Microscopy Center for Materials Research (EMCMR) at Argonne National Laboratory.

REFERENCES

[1] M.M. Thackeray, C. Wolverton, E.D. Isaacs, Electrical energy storage for transportation-approaching the limits of, and going beyond, lithium-ion batteries, *Energ Environ Sci,* 5 (2012) 7854-7863.

[2] Z. Yang, J. Zhang, M.C.W. Kintner-Meyer, X. Lu, D. Choi, J.P. Lemmon, J. Liu, Electrochemical Energy Storage for Green Grid, *Chemical Reviews,* 111 (2011) 3577-3613.

[3] M.S. Whittingham, Ultimate limits to intercalation reactions for lithium batteries, *Chemical Reviews,* (2014).

[4] P.G. Bruce, S.A. Freunberger, L.J. Hardwick, J.-M. Tarascon, Li-O2 and Li-S batteries with high energy storage, *Nat Mater,* 11 (2012) 19-29.

[5] A. Manthiram, Y. Fu, S.-H. Chung, C. Zu, Y.-S. Su, Rechargeable Lithium–Sulfur Batteries, *Chemical Reviews,* 114 (2014) 11751-11787.

[6] J. Lu, L. Li, J.-B. Park, Y.-K. Sun, F. Wu, K. Amine, Aprotic and Aqueous Li–O2 Batteries, Chemical Reviews, 114 (2014) 5611-5640.

[7] N. Yabuuchi, K. Kubota, M. Dahbi, S. Komaba, Research Development on Sodium-Ion Batteries, *Chemical Reviews,* (2014).

[8] S.-T. Myung, K. Amine, Y.-K. Sun, Surface modification of cathode materials from nano- to microscale for rechargeable lithium-ion batteries, *J Mater Chem,* 20 (2010) 7074-7095.

[9] X. Meng, X.-Q. Yang, X. Sun, Emerging Applications of Atomic Layer Deposition for Lithium-Ion Battery Studies, *Advanced Materials,* 24 (2012) 3589-3615.

[10] T. Suntola, J. Antson, Method for producing compound thin films, *United States Patent,* (1977) 4,058,430.

[11] G.N. Parsons, J.W. Elam, S.M. George, S. Haukka, H. Jeon, W.M.M. Kessels, M. Leskelä, P. Poodt, M. Ritala, S.M. Rossnagel, History of atomic layer deposition and its relationship with the American VacuumSociety, *Journal of Vacuum Science & Technology A,* 31 (2013) 050818.

[12] S.M. George, Atomic Layer Deposition: An Overview, *Chemical Reviews,* 110 (2010) 111-131.

[13] M.D. Groner, F.H. Fabreguette, J.W. Elam, S.M. George, Low-Temperature Al2O3 Atomic Layer Deposition, *Chem Mater,* 16 (2004) 639-645.

[14] X.B. Meng, J. Liu, X.F. Li, M.N. Banis, J.L. Yang, R.Y. Li, X.L. Sun, Atomic layer deposited $Li_4Ti_5O_{12}$ on nitrogen-doped carbon nanotubes, *Rsc Adv,* 3 (2013) 7285-7288.

[15] V. Miikkulainen, M. Leskelä, M. Ritala, R.L. Puurunen, Crystallinity of inorganic films grown by atomic layer deposition: Overview and general trends, *Journal of Applied Physics,* 113 (2013) 021301.

[16] C. Detavernier, J. Dendooven, S.P. Sree, K.F. Ludwig, J.A. Martens, Tailoring nanoporous materials by atomic layer deposition, *Chem Soc Rev,* 40 (2011) 5242-5253.

[17] C. Marichy, M. Bechelany, N. Pinna, Atomic Layer Deposition of Nanostructured Materials for Energy and Environmental Applications, *Adv Mater,* 24 (2012) 1017-1032.

[18] M. Knez, K. Nielsch, L. Niinistö, Synthesis and Surface Engineering of Complex Nanostructures by Atomic Layer Deposition, *Adv Mater,* 19 (2007) 3425-3438.

[19] T. Suntola, Atomic layer epitaxy, *Materials Science Reports*, 4 (1989) 261-312.

[20] O. Sneh, R.B. Clark-Phelps, A.R. Londergan, J. Winkler, T.E. Seidel, Thin film atomic layer deposition equipment for semiconductor processing, *Thin Solid Films,* 402 (2002) 248-261.

[21] S.A. Skoog, J.W. Elam, R.J. Narayan, Atomic layer deposition: medical and biological applications, *International Materials Reviews,* 58 (2013) 113-129.

[22] B.J. O'Neill, D.H.K. Jackson, J. Lee, C. Canlas, P.C. Stair, C.L. Marshall, J.W. Elam, T.F. Kuech, J.A. Dumesic, G.W. Huber, Catalyst Design with Atomic Layer Deposition, *ACS Catalysis,* 5 (2015) 1804-1825.

[23] X. Du, S.M. George, Thickness dependence of sensor response for CO gas sensing by tin oxide films grown using atomic layer deposition, *Sensors and Actuators B: Chemical,* 135 (2008) 152-160.

[24] L. Niinistö, M. Nieminen, J. Päiväsaari, J. Niinistö, M. Putkonen, M. Nieminen, Advanced electronic and optoelectronic materials by Atomic Layer Deposition: An overview with special emphasis on recent progress in processing of high-k dielectrics and other oxide materials, *physica status solidi (a),* 201 (2004) 1443-1452.

[25] T.M. Mayer, J.W. Elam, S.M. George, P.G. Kotula, R.S. Goeke, Atomic-layer deposition of wear-resistant coatings for microelectromechanical devices, *Appl Phys Lett,* 82 (2003) 2883-2885.

[26] H.C.M. Knoops, M.E. Donders, L. Baggetto, M.C.M. van de Sanden, P.H.L. Notten, W.M.M. Kessels, Atomic Layer Deposition for All-Solid-State 3D-Integrated Batteries, *Ecs Transactions,* 25 (2009) 333-344.

[27] Q. Peng, J.S. Lewis, P.G. Hoertz, J.T. Glass, G.N. Parsons, Atomic layer deposition for electrochemical energy generation and storage systems, *J Vac Sci Technol A,* 30 (2012).

[28] W. Niu, X. Li, S.K. Karuturi, D.W. Fam, H. Fan, S. Shrestha, L.H. Wong, A.I. Tok, Applications of atomic layer deposition in solar cells, *Nanotechnology,* 26 (2015) 064001.

[29] X. Meng, Towards high-energy and durable lithium-ion batteries via atomic layer deposition: elegantly atomic-scale material design and surface modification, *Nanotechnology,* 26 (2015) 020501.

[30] N.P. Dasgupta, X. Meng, J.W. Elam, A.B.F. Martinson, Atomic Layer Deposition of Metal Sulfide Materials, *Accounts of Chemical Research,* 48 (2015) 341-348.

[31] J. Hämäläinen, M. Ritala, M. Leskelä, Atomic Layer Deposition of Noble Metals and Their Oxides, *Chem Mater,* 26 (2014) 786-801.

[32] R.L. Puurunen, Surface chemistry of atomic layer deposition: A case study for the trimethylaluminum/water process, *Journal of Applied Physics,* 97 (2005) 121301.

[33] H. Kim, Atomic layer deposition of metal and nitride thin films: Current research efforts and applications for semiconductor device processing, *Journal of Vacuum Science & Technology* B, 21 (2003) 2231-2261.

[34] M. Putkonen, A. Szeghalmi, E. Pippel, M. Knez, Atomic layer deposition of metal fluorides through oxide chemistry, *J Mater Chem,* 21 (2011) 14461-14465.

[35] H. Kim, H.B.R. Lee, W.J. Maeng, Applications of atomic layer deposition to nanofabrication and emerging nanodevices, *Thin Solid Films,* 517 (2009) 2563-2580.

[36] X.B. Meng, X.Q. Yang, X.L. Sun, Emerging Applications of Atomic Layer Deposition for Lithium-Ion Battery Studies, *Adv Mater,* 24 (2012) 3589-3615.

[37] X. Meng, J.A. Libera, T.T. Fister, H. Zhou, J.K. Hedlund, P. Fenter, J.W. Elam, Atomic Layer Deposition of Gallium Sulfide Films Using Hexakis (dimethylamido)digallium and Hydrogen Sulfide, *Chem Mater,* 26 (2014) 1029-1039.

[38] R. Lupták, J.M.J. Lopes, S. Lenk, B. Holländer, E.D. Özben, A.T. Tiedemann, M. Schnee, J. Schubert, S. Habicht, S. Feste, S. Mantl, U. Breuer, A. Besmehn, P.K. Baumann, M. Heuken, Atomic layer deposition of HfO2 and Al2O3 layers on 300 mm Si wafers for gate stack technology, *Journal of Vacuum Science & Technology* B, 29 (2011) 01A301.

[39] C. Jan, J. Bielefeld, M. Buehler, V. Chikamane, K. Fischer, T. Hepburn, A. Jain, J. Jeong, T. Kielty, S. Kook, 2003 IEEE *Interconnect Technology Conference*2003.

[40] J.W. Elam, D. Routkevitch, P.P. Mardilovich, S.M. George, Conformal Coating on Ultrahigh-Aspect-Ratio Nanopores of Anodic Alumina by Atomic Layer Deposition, *Chem Mater,* 15 (2003) 3507-3517.

[41] X.B. Meng, D.S. Geng, J.A. Liu, M.N. Banis, Y. Zhang, R.Y. Li, X.L. Sun, Non-Aqueous Approach to Synthesize Amorphous/Crystalline Metal Oxide-Graphene Nanosheet Hybrid Composites, *J Phys Chem C,* 114 (2010) 18330-18337.

[42] X.Y. Chen, H.L. Zhu, Y.C. Chen, Y.Y. Shang, A.Y. Cao, L.B. Hu, G.W. Rubloff, MWCNT/V2O5 Core/Shell Sponge for High Areal Capacity and Power Density Li-Ion Cathodes, *Acs Nano,* 6 (2012) 7948-7955.

[43] X. Meng, M. Ionescu, M. Banis, Y. Zhong, H. Liu, Y. Zhang, S. Sun, R. Li, X. Sun, Heterostructural coaxial nanotubes of CNT@Fe2O3 via atomic layer deposition: effects of surface functionalization and nitrogen-doping, *J Nanopart Res,* 13 (2011) 1207-1218.

[44] A. Brzezinski, Y.-C. Chen, P. Wiltzius, P.V. Braun, Complex three-dimensional conformal surfaces formed by atomic layer deposition: computation and experimental verification, *J Mater Chem,* 19 (2009) 9126-9130.

[45] J.S. King, E. Graugnard, C.J. Summers, TiO2 Inverse Opals Fabricated Using Low-Temperature Atomic Layer Deposition, *Adv Mater,* 17 (2005) 1010-1013.

[46] G. Triani, J.A. Campbell, P.J. Evans, J. Davis, B.A. Latella, R.P. Burford, Low temperature atomic layer deposition of titania thin films, *Thin Solid Films,* 518 (2010) 3182-3189.

[47] E. Guziewicz, I.A. Kowalik, M. Godlewski, K. Kopalko, V. Osinniy, A. Wójcik, S. Yatsunenko, E. Łusakowska, W. Paszkowicz, M. Guziewicz, Extremely low temperature growth of ZnO by atomic layer deposition, *Journal of Applied Physics,* 103 (2008) 033515.

[48] J. Heo, A.S. Hock, R.G. Gordon, Low Temperature Atomic Layer Deposition of Tin Oxide, *Chem Mater,* 22 (2010) 4964-4973.

[49] M. Putkonen, L. Niinistö, Atomic layer deposition of B2O3 thin films at room temperature, *Thin Solid Films,* 514 (2006) 145-149.

[50] J.W. Klaus, S.M. George, Atomic layer deposition of SiO2 at room temperature using NH3-catalyzed sequential surface reactions, *Surface Science,* 447 (2000) 81-90.

[51] Q. Xu, J. Yang, J.W. Dai, Y. Yang, X.Q. Chen, Y. Wang, Hydrophilization of porous polypropylene membranes by atomic layer deposition of TiO2 for simultaneously improved permeability and selectivity, *J Membrane Sci,* 448 (2013) 215-222.

[52] H. Chen, Q. Lin, Q. Xu, Y. Yang, Z.P. Shao, Y. Wang, Plasma activation and atomic layer deposition of TiO2 on polypropylene membranes for improved performances of lithium-ion batteries, *J Membrane Sci,* 458 (2014) 217-224.

[53] E. Pomerantseva, K. Gerasopoulos, M. Gnerlich, P. Odenwald, J. Culver, R. Ghodssi, Using Tobacco mosaic virus template for the fabrication of three-dimensional hierarchical microbattery electrodes with high energy and high power density, *Proc Spie,* 8820 (2013).

[54] K. Gerasopoulos, X.L. Chen, J. Culver, C.S. Wang, R. Ghodssi, Self-assembled Ni/TiO2 nanocomposite anodes synthesized via electroless plating and atomic layer deposition on biological scaffolds, *Chem Commun,* 46 (2010) 7349-7351.

[55] X.B. Meng, M.N. Banis, D.S. Geng, X.F. Li, Y. Zhang, R.Y. Li, H. Abou-Rachid, X.L. Sun, Controllable atomic layer deposition of one-dimensional nanotubular TiO2, *Appl Surf Sci,* 266 (2013) 132-140.

[56] X.B. Meng, D.S. Geng, J.A. Liu, R.Y. Li, X.L. Sun, Controllable synthesis of graphene-based titanium dioxide nanocomposites by atomic layer deposition, *Nanotechnology,* 22 (2011).

[57] J. Leem, I. Park, Y. Li, W. Zhou, Z. Jin, S. Shin, Y.-S. Min, Role of HCl in Atomic Layer Deposition of TiO2 Thin Films from Titanium Tetrachloride and Water, *Bull. Korean Chem. Soc,* 35 (2014) 1195.

[58] X. Meng, Y. Zhong, Y. Sun, M.N. Banis, R. Li, X. Sun, Nitrogen-doped carbon nanotubes coated by atomic layer deposited SnO2 with controlled morphology and phase, *Carbon,* 49 (2011) 1133-1144.

[59] K. Kukli, M. Ritala, M. Schuisky, M. Leskelä, T. Sajavaara, J. Keinonen, T. Uustare, A. Hårsta, Atomic Layer Deposition of Titanium Oxide from TiI4 and H2O2, *Chemical Vapor Deposition,* 6 (2000) 303-310.

[60] S.M. Prokes, M.B. Katz, M.E. Twigg, Growth of crystalline Al2O3 via thermal atomic layer deposition: Nanomaterial phase stabilization, *Apl Mater,* 2 (2014).

[61] S. Goriparti, E. Miele, F. De Angelis, E. Di Fabrizio, R. Proietti Zaccaria, C. Capiglia, Review on recent progress of nanostructured anode materials for Li-ion batteries, *J Power Sources,* 257 (2014) 421-443.

[62] C. Liang, M.X. Gao, H.G. Pan, Y.F. Liu, M. Yan, Lithium alloys and metal oxides as high-capacity anode materials for lithium-ion batteries, *J Alloy Compd,* 575 (2013) 246-256.

[63] M.V. Reddy, G.V. Subba Rao, B.V.R. Chowdari, Metal Oxides and Oxysalts as Anode Materials for Li Ion Batteries, *Chemical Reviews,* 113 (2013) 5364-5457.

[64] X. Rui, H. Tan, Q. Yan, Nanostructured metal sulfides for energy storage, *Nanoscale,* 6 (2014) 9889-9924.

[65] J. Cabana, L. Monconduit, D. Larcher, M.R. Palacín, Beyond Intercalation-Based Li-Ion Batteries: The State of the Art and Challenges of Electrode Materials Reacting Through Conversion Reactions, *Adv Mater,* 22 (2010) E170-E192.

[66] L.J. Fu, H. Liu, C. Li, Y.P. Wu, E. Rahm, R. Holze, H.Q. Wu, Surface modifications of electrode materials for lithium ion batteries, *Solid State Sciences,* 8 (2006) 113-128.

[67] Y.S. Jung, A.S. Cavanagh, L.A. Riley, S.H. Kang, A.C. Dillon, M.D. Groner, S.M. George, S.H. Lee, Ultrathin Direct Atomic Layer Deposition on Composite Electrodes for Highly Durable and Safe Li-Ion Batteries, *Adv Mater,* 22 (2010) 2172-+.

[68] M.L. Lee, C.Y. Su, Y.H. Lin, S.C. Liao, J.M. Chen, T.P. Perng, J.W. Yeh, H.C. Shih, Atomic layer deposition of TiO2 on negative electrode for lithium ion batteries, *J Power Sources,* 244 (2013) 410-416.

[69] H.Y. Wang, F.M. Wang, Electrochemical investigation of an artificial solid electrolyte interface for improving the cycle-ability of lithium ion batteries using an atomic layer deposition on a graphite electrode, *J Power Sources,* 233 (2013) 1-5.

[70] I. Lahiri, S.M. Oh, J.Y. Hwang, C. Kang, M. Choi, H. Jeon, R. Banerjee, Y.K. Sun, W. Choi, Ultrathin alumina-coated carbon nanotubes as an anode for high capacity Li-ion batteries, *J Mater Chem,* 21 (2011) 13621-13626.

[71] A.J. Loebl, C.J. Oldham, C.K. Devine, B. Gong, S.E. Atanasov, G.N. Parsons, P.S. Fedkiw, Solid Electrolyte Interphase on Lithium-Ion Carbon Nanofiber Electrodes by Atomic and Molecular Layer Deposition, *J Electrochem Soc,* 160 (2013) A1971-A1978.

[72] D. Molina Piper, S.-B. Son, J.J. Travis, Y. Lee, S.S. Han, S.C. Kim, K.H. Oh, S.M. George, S.-H. Lee, C. Ban, Mitigating irreversible capacity losses from carbon agents via surface modification, *J Power Sources,* 275 (2015) 605-611.

[73] M.Q. Snyder, S.A. Trebukhova, B. Ravdel, M.C. Wheeler, J. DiCarlo, C.P. Tripp, W.J. DeSisto, Synthesis and characterization of atomic layer deposited titanium nitride thin films on lithium titanate spinel powder as a lithium-ion battery anode, *J Power Sources,* 165 (2007) 379-385.

[74] D. Ahn, X.C. Xiao, Extended lithium titanate cycling potential window with near zero capacity loss, *Electrochem Commun,* 13 (2011) 796-799.

[75] J. Liu, X.F. Li, M. Cai, R.Y. Li, X.L. Sun, Ultrathin atomic layer deposited ZrO2 coating to enhance the electrochemical performance of Li4Ti5O12 as an anode material, *Electrochim Acta,* 93 (2013) 195-201.

[76] A.N. Dey, Electrochemical Alloying of Lithium in Organic Electrolytes, *J Electrochem Soc,* 118 (1971) 1547-1549.

[77] C.-M. Park, J.-H. Kim, H. Kim, H.-J. Sohn, Li-alloy based anode materials for Li secondary batteries, *Chem Soc Rev,* 39 (2010) 3115-3141.

[78] X.C. Xiao, P. Lu, D. Ahn, Ultrathin Multifunctional Oxide Coatings for Lithium Ion Batteries, *Adv Mater,* 23 (2011) 3911-+.

[79] Y. He, X.Q. Yu, Y.H. Wang, H. Li, X.J. Huang, Alumina-Coated Patterned Amorphous Silicon as the Anode for a Lithium-Ion Battery with High Coulombic Efficiency, *Adv Mater,* 23 (2011) 4938-4941.

[80] E.L. Memarzadeh, W.P. Kalisvaart, A. Kohandehghan, B. Zahiri, C.M.B. Holt, D. Mitlin, Silicon nanowire core aluminum shell coaxial nanocomposites for lithium ion battery anodes grown with and without a TiN interlayer, *J Mater Chem,* 22 (2012) 6655-6668.

[81] A. Kohandehghan, P. Kalisvaart, K. Cui, M. Kupsta, E. Memarzadeh, D. Mitlin, Silicon nanowire lithium-ion battery anodes with ALD deposited TiN coatings demonstrate a major improvement in cycling performance, *J Mater Chem A,* 1 (2013) 12850-12861.

[82] E. Memarzadeh Lotfabad, P. Kalisvaart, K. Cui, A. Kohandehghan, M. Kupsta, B. Olsen, D. Mitlin, ALD TiO2 coated silicon nanowires for lithium ion battery anodes with enhanced cycling stability and coulombic efficiency, *Phys Chem Chem Phys,* 15 (2013) 13646-13657.

[83] E.M. Lotfabad, P. Kalisvaart, A. Kohandehghan, K. Cui, M. Kupsta, B. Farbod, D. Mitlin, Si nanotubes ALD coated with TiO2, TiN or Al2O3 as high performance lithium ion battery anodes, *J Mater Chem* A, 2 (2014) 2504-2516.

[84] Y. Li, Y. Sun, G. Xu, Y. Lu, S. Zhang, L. Xue, J.S. Jur, X. Zhang, Tuning electrochemical performance of Si-based anodes for lithium-ion batteries by employing atomic layer deposition alumina coating, *J Mater Chem* A, 2 (2014) 11417-11425.

[85] J.G. Ren, Q.H. Wu, G. Hong, W.J. Zhang, H.M. Wu, K. Amine, J.B. Yang, S.T. Lee, Silicon-Graphene Composite Anodes for High-Energy Lithium Batteries, *Energy Technol-Ger,* 1 (2013) 77-84.

[86] J.C. Li, X.C. Xiao, Y.T. Cheng, M.W. Verbrugge, Atomic Layered Coating Enabling Ultrafast Surface Kinetics at Silicon Electrodes in Lithium Ion Batteries, *J Phys Chem Lett,* 4 (2013) 3387-3391.

[87] J.C. Ye, Y.H. An, T.W. Heo, M.M. Biener, R.J. Nikolic, M. Tang, H. Jiang, Y.M. Wang, Enhanced lithiation and fracture behavior of silicon mesoscale pillars via atomic layer coatings and geometry design, *J Power Sources,* 248 (2014) 447-456.

[88] M.N. Obrovac, V.L. Chevrier, Alloy Negative Electrodes for Li-Ion Batteries, *Chemical Reviews,* 114 (2014) 11444-11502.

[89] D.N. Wang, J.L. Yang, J. Liu, X.F. Li, R.Y. Li, M. Cai, T.K. Sham, X.L. Sun, Atomic layer deposited coatings to significantly stabilize anodes for Li ion batteries: effects of coating thickness and the size of anode particles, *J Mater Chem* A, 2 (2014) 2306-2312.

[90] T.Q. Nguyen, A.K. Thapa, V.K. Vendra, J.B. Jasinski, G.U. Sumanasekera, M.K. Sunkara, High rate capacity retention of binder-free, tin oxide nanowire arrays using thin titania and alumina coatings, *Rsc Adv,* 4 (2014) 3312-3317.

[91] Y. Liu, N.S. Hudak, D.L. Huber, S.J. Limmer, J.P. Sullivan, J.Y. Huang, In Situ Transmission Electron Microscopy Observation of Pulverization of Aluminum Nanowires and Evolution of the Thin Surface Al2O3 Layers during Lithiation-Delithiation Cycles, *Nano Lett,* 11 (2011) 4188-4194.

[92] Z.-W. Fu, F. Huang, Y. Zhang, Y. Chu, Q.-Z. Qin The Electrochemical Reaction of Zinc Oxide Thin Films with Lithium, *J Electrochem Soc,* 150 (2003) A714-A720.

[93] J.H. Lee, M.H. Hon, Y.W. Chung, I.C. Leu, The effect of TiO2 coating on the electrochemical performance of ZnO nanorod as the anode material for lithium-ion battery, *Appl Phys a-Mater,* 102 (2011) 545-550.

[94] M. Yu, A. Wang, Y. Wang, C. Li, G. Shi, An alumina stabilized ZnO-graphene anode for lithium ion batteries via atomic layer deposition, *Nanoscale,* 6 (2014) 11419-11424.

[95] E. Kang, Y.S. Jung, A.S. Cavanagh, G.H. Kim, S.M. George, A.C. Dillon, J.K. Kim, J. Lee, Fe3O4 Nanoparticles Confined in Mesocellular Carbon Foam for High

Performance Anode Materials for Lithium-Ion Batteries, *Adv Funct Mater,* 21 (2011) 2430-2438.

[96] K. Mizushima, P.C. Jones, P.J. Wiseman, J.B. Goodenough, $LixCoO_2$ ($0<x\leq1$): A new cathode material for batteries of high energy density, *Materials Research Bulletin,* 15 (1980) 783-789.

[97] M.S. Whittingham, Lithium Batteries and Cathode Materials, *Chemical Reviews,* 104 (2004) 4271-4302.

[98] M. Hu, X. Pang, Z. Zhou, Recent progress in high-voltage lithium ion batteries, *J Power Sources,* 237 (2013) 229-242.

[99] T.-F. Yi, Y.-R. Zhu, X.-D. Zhu, J. Shu, C.-B. Yue, A.-N. Zhou, A review of recent developments in the surface modification of LiMn2O4 as cathode material of power lithium-ion battery, *Ionics,* 15 (2009) 779-784.

[100] O.K. Park, Y. Cho, S. Lee, H.-C. Yoo, H.-K. Song, J. Cho, Who will drive electric vehicles, olivine or spinel?, *Energ Environ Sci*, 4 (2011) 1621-1633.

[101] J. Cho, G. Kim, Enhancement of Thermal Stability of LiCoO2 by LiMn2 O 4 Coating, *Electrochemical and Solid-State Letters,* 2 (1999) 253-255.

[102] J. Cho, C.S. Kim, S.I. Yoo, Improvement of Structural Stability of LiCoO2 Cathode during Electrochemical Cycling by Sol-Gel Coating of SnO2, *Electrochemical and Solid-State Letters,* 3 (2000) 362-365.

[103] J. Cho, Y.J. Kim, B. Park, Novel LiCoO2 Cathode Material with Al2O3 Coating for a Li Ion Cell, *Chem Mater,* 12 (2000) 3788-3791.

[104] J. Cho, Y.J. Kim, B. Park, LiCoO2 Cathode Material That Does Not Show a Phase Transition from Hexagonal to Monoclinic Phase, *J Electrochem Soc,* 148 (2001) A1110-A1115.

[105] J. Cho, Y.-W. Kim, B. Kim, J.-G. Lee, B. Park, A Breakthrough in the Safety of Lithium Secondary Batteries by Coating the Cathode Material with AlPO4 Nanoparticles, *Angewandte Chemie International Edition,* 42 (2003) 1618-1621.

[106] Y.S. Jung, A.S. Cavanagh, A.C. Dillon, M.D. Groner, S.M. George, S.H. Lee, Enhanced Stability of LiCoO2 Cathodes in Lithium-Ion Batteries Using Surface Modification by Atomic Layer Deposition, *J Electrochem Soc,* 157 (2010) A75-A81.

[107] I.D. Scott, Y.S. Jung, A.S. Cavanagh, Y.F. An, A.C. Dillon, S.M. George, S.H. Lee, Ultrathin Coatings on Nano-LiCoO2 for Li-Ion Vehicular Applications, *Nano Lett,* 11 (2011) 414-418.

[108] J.T. Lee, F.M. Wang, C.S. Cheng, C.C. Li, C.H. Lin, Low-temperature atomic layer deposited Al2O3 thin film on layer structure cathode for enhanced cycleability in lithium-ion batteries, *Electrochim Acta,* 55 (2010) 4002-4006.

[109] H.M. Cheng, F.M. Wang, J.P. Chu, R. Santhanam, J. Rick, S.C. Lo, Enhanced Cycleabity in Lithium Ion Batteries: Resulting from Atomic Layer Depostion of Al2O3 or TiO2 on LiCoO2 Electrodes, *J Phys Chem C,* 116 (2012) 7629-7637.

[110] X.F. Li, J. Liu, X.B. Meng, Y.J. Tang, M.N. Banis, J.L. Yang, Y.H. Hu, R.Y. Li, M. Cai, X.L. Sun, Significant impact on cathode performance of lithium-ion batteries by precisely controlled metal oxide nanocoatings via atomic layer deposition, *J Power Sources,* 247 (2014) 57-69.

[111] Y.S. Jung, P. Lu, A.S. Cavanagh, C. Ban, G.H. Kim, S.H. Lee, S.M. George, S.J. Harris, A.C. Dillon, Unexpected Improved Performance of ALD Coated LiCoO2/Graphite Li-Ion Batteries, *Adv Energy Mater,* 3 (2013) 213-219.

[112] J.H. Woo, J.J. Travis, S.M. George, S.-H. Lee, Utilization of Al2O3 Atomic Layer Deposition for Li Ion Pathways in Solid State Li Batteries, *J Electrochem Soc,* 162 (2015) A344-A349.

[113] J.S. Park, A.U. Mane, J.W. Elam, J.R. Croy, Amorphous Metal Fluoride Passivation Coatings Prepared by Atomic Layer Deposition on LiCoO2 for Li-Ion Batteries, *Chem Mater,* 27 (2015) 1917-1920.

[114] J.W. Kim, J.J. Travis, E.Y. Hu, K.W. Nam, S.C. Kim, C.S. Kang, J.H. Woo, X.Q. Yang, S.M. George, K.H. Oh, S.J. Cho, S.H. Lee, Unexpected high power performance of atomic layer deposition coated Li[Ni1/3Mn1/3Co1/3]O-2 cathodes, *J Power Sources,* 254 (2014) 190-197.

[115] X.F. Li, J. Liu, M.N. Banis, A. Lushington, R.Y. Li, M. Cai, X.L. Sun, Atomic layer deposition of solid-state electrolyte coated cathode materials with superior high-voltage cycling behavior for lithium ion battery application, *Energ Environ Sci,* 7 (2014) 768-778.

[116] J.Z. Kong, C. Ren, G.A. Tai, X. Zhang, A.D. Li, D. Wu, H. Li, F. Zhou, Ultrathin ZnO coating for improved electrochemical performance of LiNi0.5Co0.2Mn0.3O2 cathode material, *J Power Sources,* 266 (2014) 433-439.

[117] Y.S. Jung, A.S. Cavanagh, Y.F. Yan, S.M. George, A. Manthiram, Effects of Atomic Layer Deposition of Al2O3 on the Li[Li0.20Mn0.54Ni0.13Co0.13]O-2 Cathode for Lithium-Ion Batteries, *J Electrochem Soc,* 158 (2011) A1298-A1302.

[118] J.Q. Zhao, S. Aziz, Y. Wang, Hierarchical functional layers on high-capacity lithium-excess cathodes for superior lithium ion batteries, *J Power Sources,* 247 (2014) 95-104.

[119] X.F. Zhang, I. Belharouak, L. Li, Y. Lei, J.W. Elam, A.M. Nie, X.Q. Chen, R.S. Yassar, R.L. Axelbaum, Structural and Electrochemical Study of Al2O3 and TiO2 Coated Li1.2Ni0.13Mn0.54Co0.13O2 Cathode Material Using ALD, *Adv Energy Mater, 3* (2013) 1299-1307.

[120] M. Bettge, Y. Li, B. Sankaran, N.D. Rago, T. Spila, R.T. Haasch, I. Petrov, D.P. Abraham, Improving high-capacity Li1.2Ni0.15Mn0.55Co0.1O2-based lithium-ion cells by modifiying the positive electrode with alumina, *J Power Sources,* 233 (2013) 346-357.

[121] I. Bloom, L. Trahey, A. Abouimrane, I. Belharouak, X.F. Zhang, Q.L. Wu, W.Q. Lu, D.P. Abraham, M. Bettge, J.W. Elam, X.B. Meng, A.K. Burrell, C.M. Ban, R. Tenent, J. Nanda, N. Dudney, Effect of interface modifications on voltage fade in $0.5Li_2MnO_3.0.5LiNi_{0.375}$ $Mn_{0.375}CO_{0.25}O_2$ cathode materials, *J Power Sources,* 249 (2014) 509-514.

[122] X. Zhang, X. Meng, J.W. Elam, I. Belharouak, Electrochemical characterization of voltage fade of Li1.2Ni0.2Mn0.6O2 cathode, *Solid State Ionics,* 268, Part B (2014) 231-235.

[123] D.S. Guan, J.A. Jeevarajan, Y. Wang, Enhanced cycleability of LiMn2O4 cathodes by atomic layer deposition of nanosized-thin Al2O3 coatings, *Nanoscale,* 3 (2011) 1465-1469.

[124] X.N. Luan, D.S. Guan, Y. Wang, Enhancing High-Rate and Elevated-Temperature Performances of Nano-Sized and Micron-Sized LiMn2O4 in Lithium-Ion Batteries with Ultrathin Surface Coatings, *J Nanosci Nanotechno,* 12 (2012) 7113-7120.

[125] D.S. Guan, Y. Wang, Ultrathin surface coatings to enhance cycling stability of LiMn2O4 cathode in lithium-ion batteries, *Ionics,* 19 (2013) 1-8.

[126] J.Q. Zhao, G.Y. Qu, J.C. Flake, Y. Wang, Low temperature preparation of crystalline ZrO2 coatings for improved elevated-temperature performances of Li-ion battery cathodes, *Chem Commun,* 48 (2012) 8108-8110.

[127] J.Q. Zhao, Y. Wang, Atomic layer deposition of epitaxial ZrO2 coating on LiMn2O4 nanoparticles for high-rate lithium ion batteries at elevated temperature, *Nano Energy,* 2 (2013) 882-889.

[128] J.Q. Zhao, Y. Wang, Surface modifications of Li-ion battery electrodes with various ultrathin amphoteric oxide coatings for enhanced cycleability, *J Solid State Electr,* 17 (2013) 1049-1058.

[129] J. Zhao, Y. Wang, Ultrathin Surface Coatings for Improved Electrochemical Performance of Lithium Ion Battery Electrodes at Elevated Temperature, *The Journal of Physical Chemistry* C, 116 (2012) 11867-11876.

[130] X. Fang, M. Ge, J. Rong, Y. Che, N. Aroonyadet, X. Wang, Y. Liu, A. Zhang, C. Zhou, Ultrathin Surface Modification by Atomic Layer Deposition on High Voltage Cathode LiNi0.5Mn1.5O4 for Lithium Ion Batteries, *Energy Technol-Ger,* 2 (2014) 159-165.

[131] J.W. Kim, D.H. Kim, D.Y. Oh, H. Lee, J.H. Kim, J.H. Lee, Y.S. Jung, Surface chemistry of LiNi0.5Mn1.5O4 particles coated by Al2O3 using atomic layer deposition for lithium-ion batteries, *J Power Sources,* 274 (2015) 1254-1262.

[132] X.C. Xiao, D. Ahn, Z.Y. Liu, J.H. Kim, P. Lu, Atomic layer coating to mitigate capacity fading associated with manganese dissolution in lithium ion batteries, *Electrochem Commun,* 32 (2013) 31-34.

[133] J.S. Park, X.B. Meng, J.W. Elam, S.Q. Hao, C. Wolverton, C. Kim, J. Cabana, Ultrathin Lithium-Ion Conducting Coatings for Increased Interfacial Stability in High Voltage Lithium-Ion Batteries, *Chem Mater,* 26 (2014) 3128-3134.

[134] B. Xiao, J. Liu, Q. Sun, B. Wang, M.N. Banis, D. Zhao, Z. Wang, R. Li, X. Cui, T.-K. Sham, X. Sun, Unravelling the Role of Electrochemically Active $FePO_4$ Coating by Atomic Layer Deposition for Increased High-Voltage Stability of $LiNi_{0.5}Mn_{1.5}O_4$ Cathode Material, *Advanced Science,* (2015) DOI: 10.1002/advs.201500022.

[135] H. Kim, J.T. Lee, D.C. Lee, A. Magasinski, W.I. Cho, G. Yushin, Plasma-Enhanced Atomic Layer Deposition of Ultrathin Oxide Coatings for Stabilized Lithium-Sulfur Batteries, *Adv Energy Mater,* 3 (2013) 1308-1315.

[136] M.P. Yu, W.J. Yuan, C. Li, J.D. Hong, G.Q. Shi, Performance enhancement of a graphene-sulfur composite as a lithium-sulfur battery electrode by coating with an ultrathin Al2O3 film via atomic layer deposition, *J Mater Chem* A, 2 (2014) 7360-7366.

[137] J. Lu, Y. Lei, K.C. Lau, X.Y. Luo, P. Du, J.G. Wen, R.S. Assary, U. Das, D.J. Miller, J.W. Elam, H.M. Albishri, D. Abd El-Hady, Y.K. Sun, L.A. Curtiss, K. Amine, A nanostructured cathode architecture for low charge overpotential in lithium-oxygen batteries, *Nat Commun,* 4 (2013).

[138] Y. Lei, J. Lu, X.Y. Luo, T.P. Wu, P. Du, X.Y. Zhang, Y. Ren, J.G. Wen, D.J. Miller, J.T. Miller, Y.K. Sun, J.W. Elam, K. Amine, Synthesis of Porous Carbon Supported

Palladium Nanoparticle Catalysts by Atomic Layer Deposition: Application for Rechargeable Lithium-O-2 Battery, *Nano Lett,* 13 (2013) 4182-4189.

[139] X.G. Han, Y. Liu, Z. Jia, Y.C. Chen, J.Y. Wan, N. Weadock, K.J. Gaskell, T. Li, L.B. Hu, Atomic-Layer-Deposition Oxide Nanoglue for Sodium Ion Batteries, *Nano Lett,* 14 (2014) 139-147.\

In: Atomic Layer Deposition (ALD)
Editor: Jeannie Valdez

ISBN: 978-1-63483-869-6

Chapter 3

Atomic Layer Deposition for Novel Nonvolatile Memory Devices

Ai-Dong Li [*]

Department of Materials Science and Engineering, College of Engineering and Applied Sciences, National Laboratory of Solid State Microstructures, Collaborative Innovation Center of Advanced Microstructures, Nanjing University, Nanjing, P. R. China

Abstract

This chapter deals with some novel nonvolatile memory devices, among which the functional layers prepared by atomic layer deposition (ALD) play an important and unique role in memory applications. It includes the charge trapping memory (CTM), resistive random access memory (RRAM), ferroelectric random access memory (FeRAM), and patterned magnetic storage, focusing on the ALD fabrication processing, microstructure, and their data storage characteristics. For CTM devices, a series of metal nanocrystal (NC)/high-*k* dielectric NC memory and new silicon-oxide-nitride-oxide-silicon (SONOS)-type CTM with excellent charge trapping capability are presented and ALD as main and flexible fabrication method is employed to deposit tunneling/blocking layer and charge trapping layer such as NCs or high-*k* composite materials. ALD-derived trilayer structure and NC RRAM cells have exhibited reliable and reproducible resistive switching behaviors. Some elements doped Hf-based oxide ferroelectric thin films have been explored by ALD for FeRAM applications. Finally, the combination of ALD-capping layer and self-assembled FePt NCs provides a new potential approach to fabricate patterned magnetic recording media with ultrahigh areal density (50Tbit/inch^2). All these indicate that ALD technique is the most suitable and necessary method for depositing the critical layers in the emerging nonvolatile memory devices with tremendous prospects.

[*] E-mail address: adli@nju.edu.cn.

INTRODUCTION

The stringent and tremendous demand for nonvolatile memory with higher density, faster speed and lower power consumption has driven the continued dimensional scaling down in Si-based integrated circuits (ICs) and computer processors [1, 2]. Over the last four decades, semiconductor-device cell-design changes and, specifically, strict shrinking, became the major impetuses for every new generation of highly integrated logic or memory devices according to famous Moore's law [3, 4]. However, with the aggressive scaling down of conventionally applied functional layers in microelectronic devices, mandatory film thicknesses are approaching their physical and dimensional limits. To overcome this hurdle, a number of new material solutions for functional layers, new innovative integrated device structures and fundamental, and new chip fabrication technology are currently being widely explored and investigated.

For memory devices, floating gate memory referred to as flash memory dominates the market of nonvolatile memory now, which is widely used in consumer electronic products such as cell phones and music players while flash-based solid-state disks (SSDs) are increasingly displacing hard disk drives as the primary storage device in laptops, desktops, and even data centers [5]. When the technology node enters the 22 nm generation, the conventional flash memory technique will also reach its physical and integration limitations and many new types of memory to replace conventional flash memories have been proposed, such as charge trapping memory (CTM), resistive random-access memory (RRAM), ferroelectric random-access memory (FeRAM), phase-change memory (PCM), and magnetic random-access memory (MRAM) [3]. Among them, CTM has a lot of advantages of compatibility with semiconductor processing, low operating voltage, and good endurance, especially for nanocrystal (NC) memory using discrete charge storage unit [6, 7]. Meanwhile RRAM is also a promising candidate for the new type nonvolatile memories because of its simple cell structure, low power consumption, superior scalability, and complementary metal-oxide semiconductor (CMOS) compatibility [8]. Early in 1990s, FeRAMs have drawn considerable attention for nonvolatile memory applications due to the advantages such as low read/write voltage, fast access time, and hard radiation [9]. The recent discovery of ferroelectricity in ultrathin Hf-based oxide thin films has spurred the new interest in its suitability for FeRAM [10]. In addition, it is of importance to develop the ultrahigh density (50 Tbit/inch^2) patterned data storage units to circumvent the superparamagnetism problem for next generation of magnetic recording technology [11].

With the continuous dimensional scaling down in semiconductor devices, now clearly entering the nanoscale node, the need for a deposition technique to produce very conformal and ultrathin films at low growth temperatures has greatly increased. Atomic layer deposition (ALD) appears to be a very timely and crucial process technique for ultra-scaled conventional and novel microelectronic devices [12]. Since 2001 the international technology roadmap for semiconductors (ITRS) regarded ALD as candidate technology preferred for microelectronics industry along with metalorganic chemical vapor deposition (MOCVD) and plasma-enhanced CVD [3], ALD has become one of the most promising and competitive film deposition approaches for microelectronics, optoelectronics, energy, and nanotechnology due to its unique advantages such as simple and precise thickness control, excellent three-dimensional (3D) conformality, large-area uniformity, and low growth temperature [13].

The strong drive force from microelectronics has triggered tremendous research efforts in ALD fabrication of new functional layers and innovative integrated steps for logic or memory devices, especially high density and 3D capacitor structures [13-15].

This chapter describes some novel nonvolatile memory such as a series of CTM devices of NC memory and silicon-oxide-nitride-oxide-silicon (SONOS)-type, RRAM, FeRAM, and patterned magnetic storage, among which the functional layers prepared by ALD play an important and unique role in memories. The ALD fabrication processing, device microstructure, and their memory characteristics have been investigated deeply. All these indicate that ALD technique is the most suitable and indispensable method for depositing the critical layers in state-of-the-art nonvolatile memory devices.

Charge Trapping Memory

The floating-gate (FG) structure, invented by Sze and Kahng at Bell Labs in 1967 [16], forms the primary technology necessary to construct flash memories. However in order to meet the demands of product miniaturization, conventional flash memory suffers from a serious reliability issue, because the floating gate of polysilicon as a charge storing layer is conductive and all charge will be lost if a leakage path appears in the very thin tunneling oxide [7]. To solve this problem, charge trapping memory (CTM) such as silicon-oxide-nitride-oxide-silicon (SONOS) has gained a lot of attention due to its low operating voltage, fast program/erase speed, good endurance and retention characteristics over the FG devices [17]. Discrete NC memory as a kind of new CTM was also first proposed by IBM in 1995 [18], and by the early 2000s researchers were already considering it to be a promising candidate for the solution of the scaling problem with high memory density due to its discrete electron storing center.

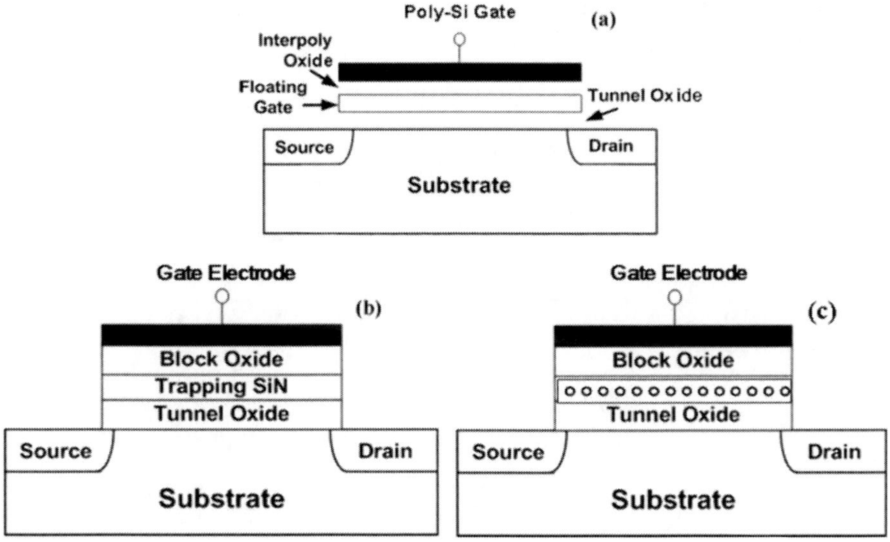

Figure 1. Schematic diagrams of (a) floating gate memory structure, (b) SONOS type-CTM and (c) nanocrystal CTM structure.

Figure 1 give the schematic diagrams of conventional FG memory, SONOS type-CTM, and novel nanocrystal CTM structure. A flash memory cell is simply a MOSFET cell, except that a polysilicon floating gate (or a silicon nitride charge trap layer) is sandwiched between a tunnel oxide and an interpolyoxide to form a charge storage layer. In nanocrystal CTM, information is stored in the NCs by injecting or removing charges, whereas in new SONOS-type CTM, high-*k* materials is utilized as charge trapping layer (CTL) for data storage.

The following section mainly presents the ALD fabrication, microstructures and memory characteristics of the nanocrystal CTM and new SONOS-type CTM devices, such as several kinds of metal NC/high-*k* dielectric NC memory cells and SONOS-type CTM using high-*k* composites as CTL.

Nanocrystal Memory

Metal Nanocrystal CTM

Metal nanocrystal (NC) CTM memory has received great attention due to its unique advantages such as high storage density, wide range choice of work function, and no multi-carrier confinement effect [19]. It consists of semiconductor substrate\tunneling layer\metal NCs\blocking layer\gate electrode. At present, the main preparation method of metal nanocrystal storage layer is high temperature annealing, which has disadvantages of poor controllability and repeatability. It is still a tough challenge to explore NCs deposition technique with uniform distribution and reliable processing.

In this context, ALD technology was chosen as the metal NC memory cell fabrication method. Pt NCs, Ir NCs, Au NCs, FePt and NCs storage layers was prepared by ALD, Ostwald ripening and self-assembly method, respectively. High-*k* thin films of Al_2O_3 and HfO_2 (or Al_2O_3) were deposited as tunneling layer and blocking layer by ALD, respectively. The effect of the processing parameters on NCs formation, arrangement, density, and structure has been investigated systematically. The impact of ALD high k tunneling/blocking layers, NC size and inter-NC distance on the storage properties of metal NC memory cells has also been studied deeply [20-23].

Table 1. Architecture and growth processing of 4 kinds of metal NC memory cells

Metal NC memory	Substrate	Tunneling layer	Storage layer	Blocking layer	Ref.
Pt	p-Si (100) (1-10 Ω·cm)	ALD Al_2O_3 (4 or 6 nm-thick)	ALD Pt (70 cycles)	ALD Al_2O_3/HfO_2 (10 or 20 nm-thick)	[20]
Ir	ibid	ALD Al_2O_3 (6 nm-thick)	ALD Ir (90 cycles)	ALD HfO_2 (20 nm-thick)	[21]
Au	ibid	ALD Al_2O_3 (3 nm-thick)	Sputtered Au (2s-20s) /RTA	Sputtered Al_2O_3 (9 nm-thick)	[22]
FePt	n-Si (100) (1-10 Ω·cm)	ALD Al_2O_3 (4 nm-thick)	Chemically synthesized FePt /self-assembly	ALD Al_2O_3 (20 nm-thick)	[23]

Table 2. ALD processing parameters of Al_2O_3, HfO_2, Pt and Ir

Material	Source and work T	Deposition T	ALD cycle
Al_2O_3	TMA* R. T. H_2O R. T	200/250/300°C	0.1s TMA /4 or 6s N_2 /0.1s H_2O /4s or 6s N_2
HfO_2	$Hf(NMe_2)_4$ 78°C H_2O R. T	300°C	0.1s $Hf(NMe_2)_4$ /4s N_2/0.1s H_2O /4s N_2
Pt	$MeCpPtMe_3$* 70°C O_2** 100 sccm	300°C	0.2s $MeCpPtMe_3$/5s N_2 /1.5s O_2/20s N_2
Ir	$Ir(acac)_3$* 200°C O_2** 40 sccm	300°C	1.5s $Ir(acac)_3$ /5s N_2 /0.7s O_2/ 5s N_2

* TMA: $Al(CH_3)_3$; acac: 2,4-pentanedionate; R. T.: room temperature.
Me: methyl; Cp: cyclopentadienyl .
** O_2: 49% O_2 + 51% N_2.

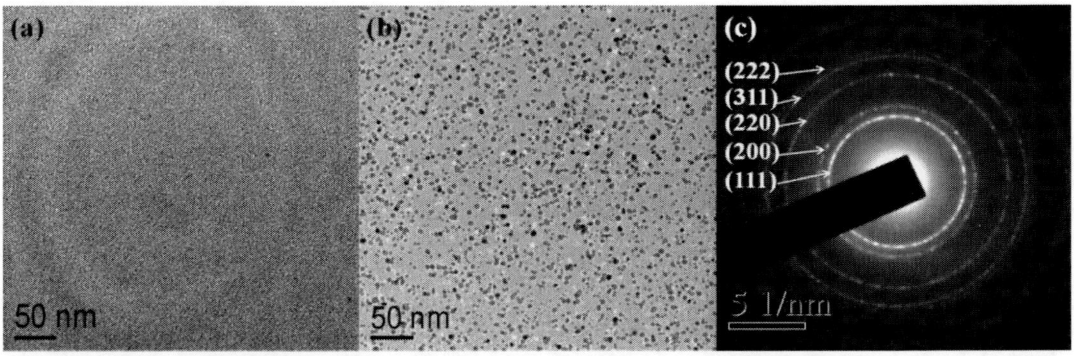

Reprinted with permission from [21]; copyright 2014, American Vacuum Society.

Figure 2. TEM images of Ir NCs after 100 ALD cycles *ex situ* (a) and *in situ* (b) deposited on ALD Al_2O_3; (c) The selected area electron diffraction pattern from *in situ* ALD Ir NCs.

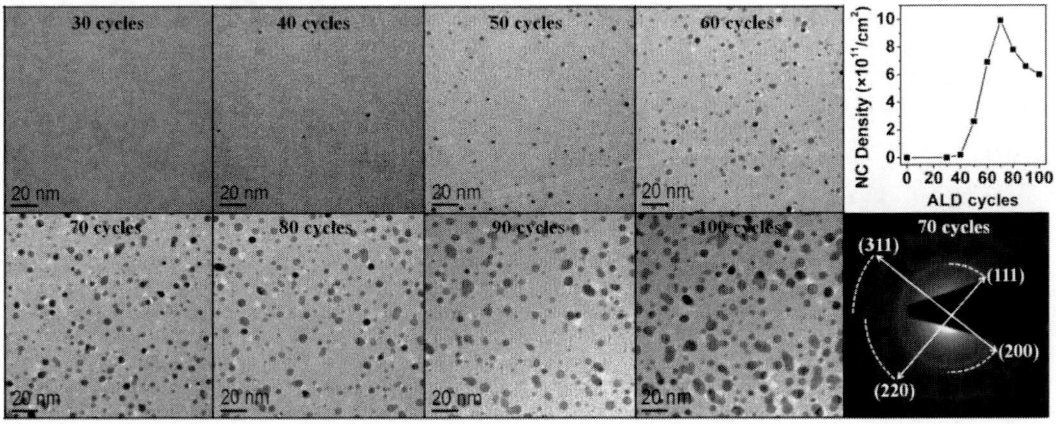

Reprinted with permission from [20]; copyright 2014, Elsevier.

Figure 3. TEM images of ALD Pt NCs deposited directly onto Al_2O_3 surface as a function of the number of ALD cycles. (Top right) Plot of NC area density extracted from TEM images. (Bottom right) The diffraction pattern from Pt NCs after 70 ALD cycles.

Table 1 lists the architecture and growth processing of 4 kinds of metal NC memory cells [20-23]. The gate electrode for metal NC memory cells was Pt prepared by sputtering method. Table 2 gives the thermal ALD processing parameters of Al_2O_3, HfO_2, Pt and Ir. ALD was performed in a commercial Picosun SUNALE^{TM} R-150B reactor using N_2 (5N) as the carrier/purge gas.

For metal NC memory, it is important to obtain metal NCs with uniform distribution and high density. The substrate surface and its pretreatment have great influence on the nucleation of ALD Ir NCs [21]. Ir cannot be directly deposited on H-terminated Si due to lack of active groups on H-Si. ALD-Al_2O_3 surfaces have various active groups by different surface pretreatment. The *ex situ* nucleation of Ir NCs on ALD Al_2O_3 is difficult because the OH group is less after the passivation of ALD-Al_2O_3 in the air. Ir NCs can be easily *in situ* formed on Al_2O_3 (Figure 2), ascribed to more active groups adsorbed on Al_2O_3 pretreated by TMA or H_2O pulse. And surface with $Al(CH_3)_2$ groups is more active than surface with OH groups, leading to the higher density of Ir NCs on Al_2O_3. The formation of *in situ* ALD Ir NCs on Al_2O_3 is similar to that of Pt NCs.

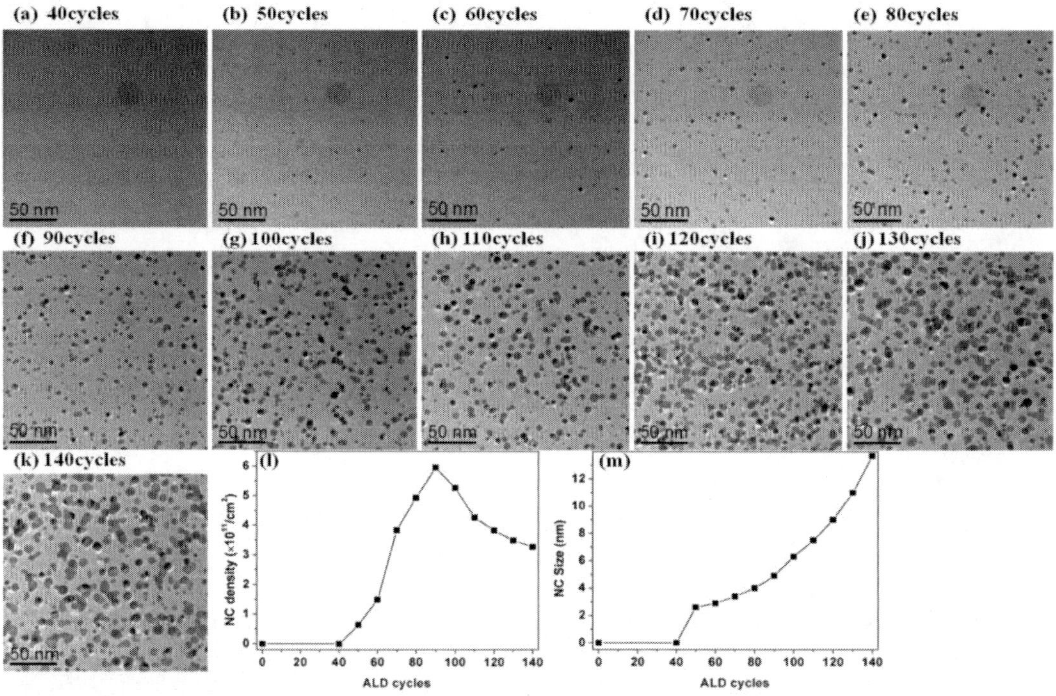

Reprinted with permission from [21]; copyright 2014, American Vacuum Society.

Figure 4. (a)–(k) TEM images of *in situ* ALD-grown Ir NCs deposited directly onto an Al_2O_3 surface as a function of the number of ALD cycles, ranging from 40 to 140 cycles. Plots of the NC (l) density and (m) average size as a function of the number of ALD cycles.

The effect of ALD cycles on Pt/Ir NCs nucleation, size, and density has been investigated systematically [20, 21], as shown in Figure 3 and 4. The formation of Pt/Ir NCs on Al_2O_3 surfaces follows the nucleation incubation model with a nucleation delay of 30~40/40~50 ALD cycles, respectively [24]. Then the density and size of Pt/Ir NCs increase with the cycle number. Merging between Pt/Ir NCs dominates the growth after 80/100 cycles, the density reduces. The maximum density of Pt NCs with average size of 3.9 nm is approximately $1.0\times10^{12}/cm^2$ at 70 cycles. The highest density of Ir NCs with the size of 4.9 nm is $0.6\times10^{12}/cm^2$ at 90 cycles. Evidently, the Pt NCs density is larger than that of Ir NCs with smaller size, implying that the nucleation in ALD Pt is much easier than that in ALD Ir. The compositional and structural analyses indicate that ALD Pt/Ir NCs are pure metallic Pt/Ir with fcc phase.

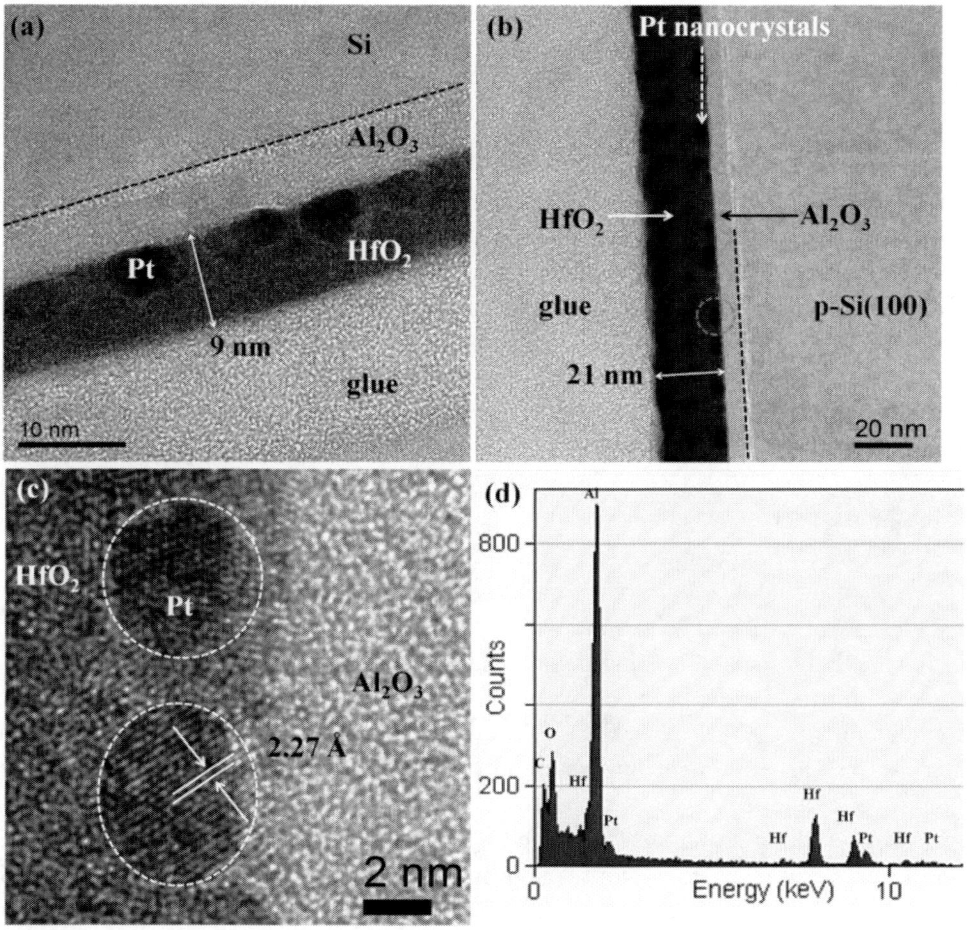

Figure 5. Cross-sectional TEM images of the p-Si\$Al_2O_3$\Pt NCs\$HfO_2$ stacking structure: (a) ALD-HfO_2 for 60 cycles, (b) ALD-HfO_2 for 180 cycles, (c) high-resolution TEM of (b); EDX spectra acquired from the red encircled area in (b).

Transmission electron microscope (TEM) together with energy dispersive x-ray (EDX) spectra was explored to characterize the cross-sectional structure of Pt NC memory cell of the p-Si\$Al_2O_3$\Pt NCs\$HfO_2$, as shown in Figure 5. The thickness of the Al_2O_3 tunneling oxide in Figure 5(a) and (b) is 6 nm, in good agreement with the nominal experimental value of 60 ALD cycles. The total thickness of the Al_2O_3\Pt NCs\180 ALD cycle-HfO_2 stacking structure is uniform and about 27 nm with amorphous Al_2O_3 and HfO_2 layers in Figure 5(b). The HRTEM image in Figure 5(c) clearly reveals Pt NCs embedded in HfO_2 layers. The (111) lattice plane of Pt NCs with lattice spacing of 2.27 Å is slightly larger than that of bulk Pt, in consistent with the above polycrystalline diffraction results. Sharp and clear interface between HfO_2 and Al_2O_3 is also confirmed, indicating no interfacial diffusion between these two oxides. The EDX spectra acquired from the red encircled area in Figure 5(b) is illustrated in Figure 5(d). The local chemical compositions around Pt NCs consist of C, O, Al, Hf and Pt, which originate from the absorbed carbon, the Al_2O_3 tunneling layer, the HfO_2 blocking layer, and the Pt NCs.

Two kinds Pt NC memory cells of Si\$Al_2O_3$\Pt NCs\$Al_2O_3$\Pt and Si\$Al_2O_3$\Pt NCs\$HfO_2$\Pt were fabricated to make comparison with memory effect [20]. The appropriate tunneling layer thickness (6 nm) and high-k blocking layer of HfO_2 are needed to achieve good storage properties. The distribution of voltage drops on tunneling and blocking layers can be easily changed by replacing Al_2O_3 blocking layer with high-k HfO_2 layer, to obtain large memory window and good retention of Pt NC memory cells. The normalized high frequency (1 MHz) C-V curves have been measured by sweeping the gate voltage from a positive bias to a negative bias and then inversely. During the backward bias sweeping, the memory capacitor undergoes a positive bias stress at the initial stage. The electrons can tunnel into the Pt NCs from the p-type Si substrate. Inversely, during the forward bias sweeping, the memory capacitor undergoes a negative bias stress at the initial stage and the holes can tunnel into the Pt NCs from the substrate.

The scanning gate voltage on MOS capacitors with Al_2O_3\Pt NCs\$HfO_2$ stacking structure changes from ±4 to ±12 V and the observed memory window increases from 0.04 to 6.6 V, as seen in Figure 6(a). The C-V characteristics of stacking structures of Al_2O_3\Pt NCs\$HfO_2$ and Al_2O_3\Pt NCs\ Al_2O_3 are compared in Figure 6(b). Sample with HfO_2 blocking layer presents a 4.0 V memory window, however sample with Al_2O_3 blocking layer has only a 0.3 V memory window at ±10 V. The sample without Pt NCs (Al_2O_3\$HfO_2$) exhibits a typical C-V curve with negligible hysteresis, indicating that the memory effect of the Pt NCs memory cells comes from the Pt NC storage instead of defect or interface state charging. Electrical measurements of Si\$Al_2O_3$\Pt NCs\$HfO_2$\Pt show better retention property with 27% charge loss after 1×10^5 s and excellent fatigue-endurance after 10^5 cycles of programming/erasing (P/E) (Figure 6 (c) and (d)).

In the two stacking structures, the blocking layer (~17 nm) is about triple thickness of tunneling layer (6 nm). The equivalent Al_2O_3 thickness of HfO_2 blocking layer is about 6 nm based on the permittivity of Al_2O_3 (~7) and HfO_2 (~20). According to the band alignment structure (Figure 7), the potential drops across the tunneling layers of p-Si\$Al_2O_3$\Pt NCs\$HfO_2$ and p-Si\$Al_2O_3$\Pt NCs\$Al_2O_3$ at applied gate voltage of 10 V are about 5.0 V and 2.5 V using Gauss's law [25], respectively. And, the electric field intensities in the tunneling layers of these two stacking structures are 8.3 MV/cm and 4.2 MV/cm, respectively. The dominant programming mechanism of MOS capacitor with p-Si\$Al_2O_3$\Pt NCs\$HfO_2$ stack layers is Fowler-Nordheim (FN) tunneling [26]. Whereas, the small electric field intensity of 4.2 MV/cm in the tunneling layer of p-Si\$Al_2O_3$\Pt NCs\$Al_2O_3$ stack structure at 10 V programming voltage makes electrons tunneling through 6 nm-thick Al_2O_3 tunneling layer more difficultly. This explains the difference of memory window of p-Si\$Al_2O_3$\Pt NCs\$HfO_2$ and p-Si\$Al_2O_3$\Pt NCs\$Al_2O_3$, as indicated in the C-V curves in Figure 5(b). Therefore the P/E mechanism can be tuned by choosing proper blocking layer materials, and the storage properties can be optimized.

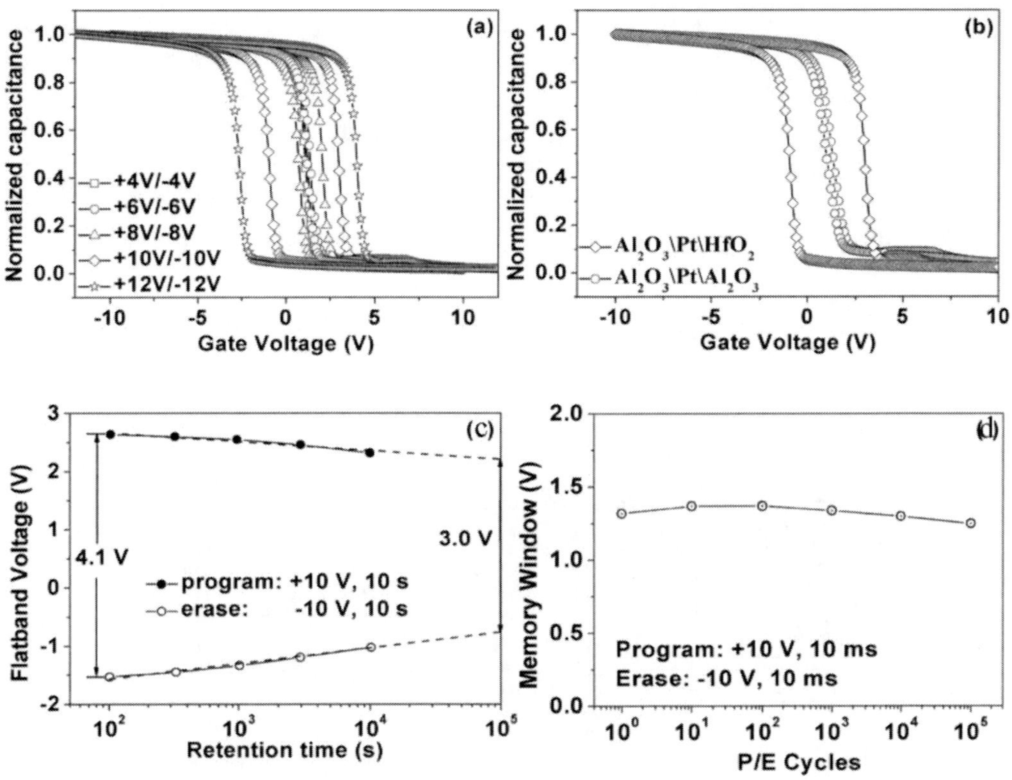

Reprinted with permission from [20]; copyright 2014, Elsevier.

Figure 6. Normalized high frequency (1 MHz) C-V curves of (a) Si\$Al_2O_3$\Pt NCs\$HfO_2$ under different sweep voltage ranges, (b) stacking structures with Al_2O_3 and HfO_2 blocking layers; Retention (c) and fatigue (d) characteristics of MOS capacitor with Al_2O_3\Pt NCs\$HfO_2$ stacking structure.

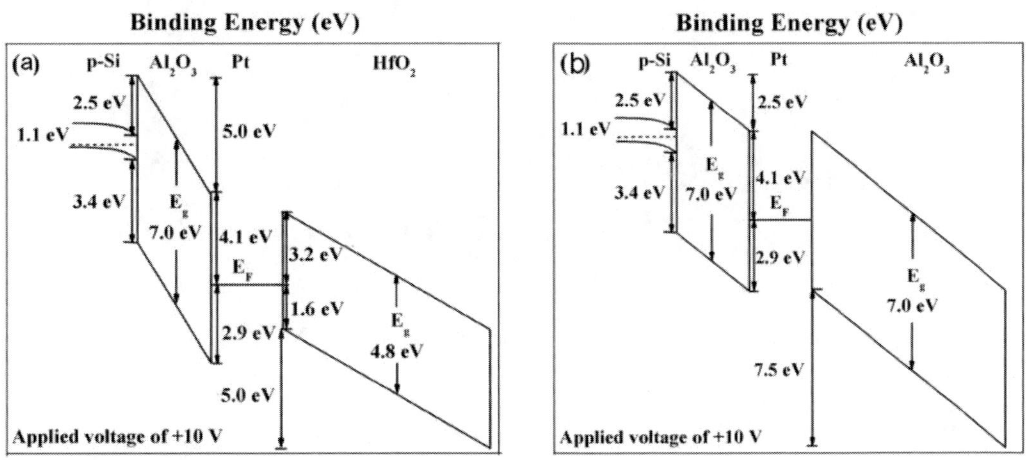

Reprinted with permission from [20]; copyright 2014, Elsevier.

Figure 7. Schematic band diagrams of stacking structures with (c) HfO_2 blocking layer and (d) Al_2O_3 blocking layer under the programming voltage of 10 V.

The memory cell of Si\$Al_2O_3$\Ir NCs\$HfO_2$ show a memory window of 4.2 V at ±10 V and ~68% retention property after 1×10^5 s. Compared to Ir NCs memory cells, Pt NCs memory cells with similar structures exhibit relatively better memory characteristics with higher storage density and retention (Table 3), which is ascribed to easier Pt nucleation on ALD-Al_2O_3 surface than Ir and deeper potential wells in Pt NCs memory cells [27].

Table 3. Comparison between Si\$Al_2O_3$\Ir NCs\$HfO_2$ and Si\$Al_2O_3$\Pt NCs\$HfO_2$

Metal NC memory	Areal density (/cm^2)	Average grain size (nm)	Memory window (V)	Retention (10^5s)	Electron/per NC (program)	Electron/per NC (erase)
Ir	0.6×10^{12}	4.9	4.2	68%	8	35
Pt	1.0×10^{12}	3.9	4.0	73%	11.7	12.6

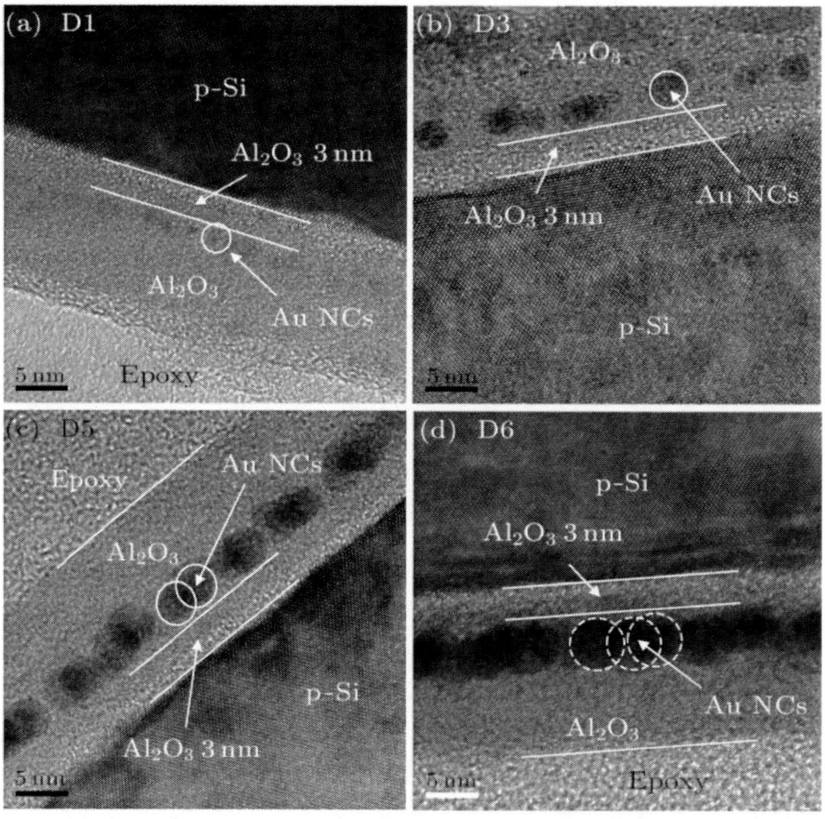

Figure 8. The cross-sectional HRTEM images of Si\$Al_2O_3$\Au NCs\$Al_2O_3$ stacking structure with original Au film of (a) 0.2 nm, (b) 0.8 nm, (c) 1.6 nm, and (d) 2.0 nm in thickness.

For memory cell of Si\$Al_2O_3$\Au NCs\$Al_2O_3$, the impact of Au NC size and inter-nanocrystal distance on the storage characteristics has been studied, as seen in Figure 8 and Table 4 [22].

ALD Al_2O_3 was used as tunneling and blocking layers. Rapid thermal annealing (RTA) process was carried out at 300°C for 10 s in a N_2 ambient to transform the ultra-thin sputtered

Au film with different thicknesses into separated Au NCs. The inter-NC distances are about 2.66 nm, 1.82 nm, and 0.87nm for D1, D3, and D5, respectively (Figure 8(a), (b) and (c)). It is obvious that the inter-NC distance decreases with the increase of the average size of Au NCs. When the average size of Au NCs is 4.65 nm, almost all Au NCs overlap with each other, as shown in Figure 8(d). Different memory windows of Au NC memory devices indicate that the charge trapping capability is dependent on the joint effect of both the size of Au NCs and the inter-NC distance. The device with a large size of Au NCs and a suitable inter-NC distance has a good charge trapping capability for D3 sample. In NC memory devices, lateral charge diffusion and vertical charge loss are the two main mechanisms for the loss of the trapped charges [28, 29]. In lateral charge diffusion, the trapped charges on NCs have a probability hopping to a neighbor site to find lower energy state via Variable range hopping (VRH) mechanism [30].Variable range hopping as the lateral charge loss mechanism is considered as the main reason for the decrease of the charge trapping capability when Au NCs grow and overlap neighbors.

Table 4. Thickness of Au films, size of NCs, NC density, and inter-NC distance of Au NCs according to the different deposition times, experimentally obtained memory window of the devices, and calculation value of the density of the stored charges

Deposition time (s)	Nominal thickness (nm)	Average Au size (nm)	Areal density (/cm^2)	Inter-NC distance (nm)	Memory window (V)	Stored charge density (/cm^2)
2	0.2	1.22±0.29	7.30±1.82	2.66±0.67	0.42	1.08
8	0.8	2.71±0.27	5.39±0.97	1.82±0.47	4.55	9.67
16	1.6	4.02±0.31	8.67±1.08	0.87±0.26	1.57	3.41
20	2.0	4.65±0.32	8.78±1.58	N/A	0.19	0.45

(Reprinted with permission from [22]; copyright 2013, IOP Publishing)

In order to obtain ordered metal NCs arrays in memory cells, a simple approach is also developed to fabricate oxide/FePt NCs/oxide composite films by a combination of chemically-synthesized FePt NCs' self-assembly and ALD [23]. Monodispersed FePt NCs with a grain size of 4.5 nm were prepared by the chemical solution method [31, 32]. The tunneling oxide Al_2O_3 layer was firstly deposited on n-Si (001) substrate by ALD, and then the FePt NCs were assembled onto Al_2O_3 layer by solution-based dip-coating. Next, a thick Al_2O_3 oxide layer was also deposited as a blocking oxide by ALD. A hexagonally arranged monolayer of well-monodispersed FePt nanocrystals has been obtained on ALD Al_2O_3 oxide, as seen in Figure 9 (a). The lattice constant of this two-dimensional pattern is about 8 nm with the high density of $1.8\times10^{12}/cm^2$. A fraction of the Fe is oxidized during annealing at 500°C for 5 min in O_2 atmosphere, and the core-shell structure is formed with fcc-$Fe_{0.75}Pt$ NC core and amorphous Fe_2O_3 shell (Figure 9 (b) and (c)). A cross-sectional TEM image of the Al_2O_3/FePt NCs/Al_2O_3 stacking structure is recorded in Figure 9(d). It can be seen that the monolayer FePt NCs are well distributed in amorphous Al_2O_3 with 4 nm-thick tunnel oxide and 20 nm-thick control oxide.

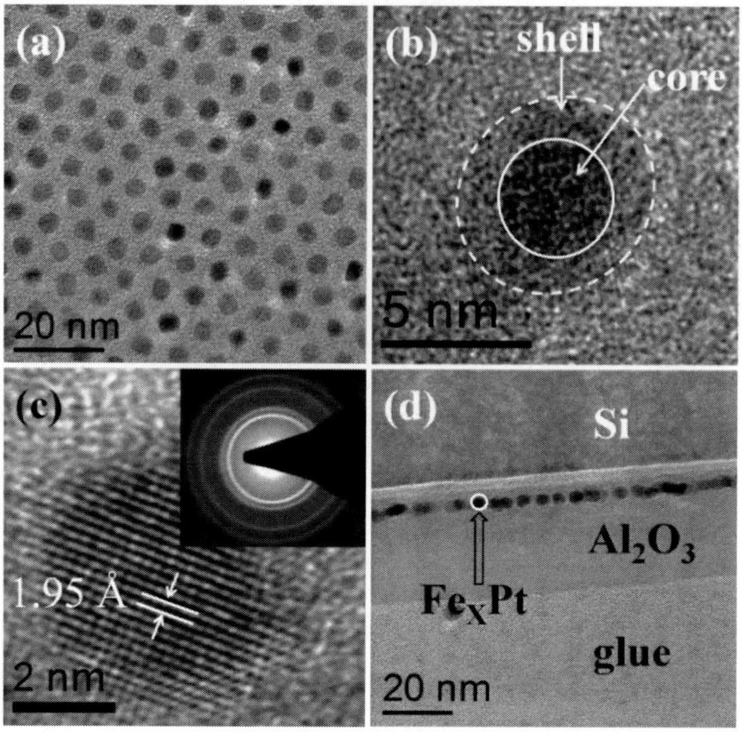

Figure 9. (a) Plan-view TEM image of assembled FePt NCs on Al_2O_3 surface without annealing; TEM images of assembled FePt NCs on Al_2O_3 surfaces after annealing at 500°C for 5 min in O_2 atmosphere: (b) HRTEM image of the core-shell structure of one FePt NC; (c) the crystalline structure of one Fe_xPt core. Inset is the diffraction pattern from FePt NCs assembly, and (d) a cross-sectional TEM image of the Si/Al_2O_3/FePt NCs/Al_2O_3 stacking structure.

The metal-oxide-Si capacitors with unanealed and annealed FePt NCs embedded into Al_2O_3 films were electrically measured, and exhibit obvious memory effects with a hysteresis memory window of 4.1 and 8.1 V at the sweeping gate voltage of ±8 V, respectively (Figure 10). The enhanced memory window of samples with annealed FePt NCs can be attributed to the existence of the Fe_2O_3 shell, which introduces additional interface and provides more trap sites for charge trapping storage [33].

High-k Dielectric Nanocrystal CTM

The traditional flash memory devices using SiO_2 as tunneling layer have suffered from some issues such as the choice between P/E speed and good data retention time due to the physical limits. To overcome these problems, high-*k* materials as the tunneling layer can provide lower tunnel barriers and smaller equivalent oxide thickness (EOT) [34]. The devices with high-*k* nanocrystals (NCs) as the charge trapping layer are also getting much attention. Compared with the conventional charge trapping layers, high-*k* dielectric films allow a higher electric field over the tunneling layer because of electric flux density continuity [35], thus improving P/E speed. Besides, high-*k* materials as blocking layer, such as Al_2O_3 film, can enhance the device performance like lower P/E voltage and scaling ability.

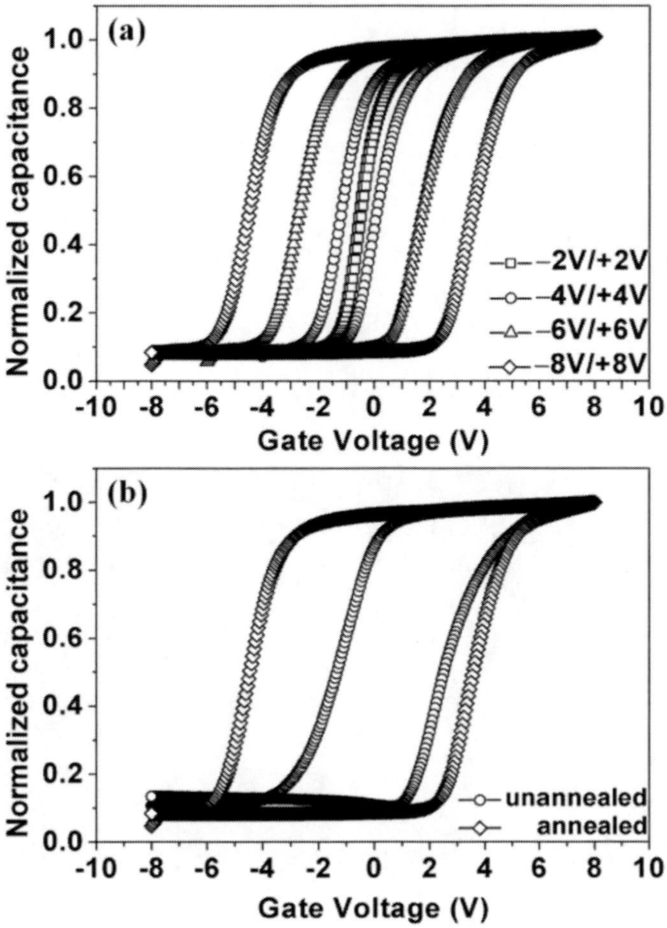

Reprinted with permission from [20]; copyright 2014, Elsevier.

Figure 10. C-V sweep of (a) MOS memory device with annealed FePt NCs under different scanning gate voltage from ±2 to ±8 V and (b) memory cells with unannealed and annealed FePt NCs at the sweeping gate voltage of ±8 V.

Table 5. Architecture and growth processing of 4 kinds of high-*k* NC memory cells

High-*k* NC memory	Tunneling layer	Storage layer	Block layer	RTA Post-annealing	Ref.
LaO	ALD Al_2O_3 (3 nm-thick)	ALD Hf-La-O (120 cycles)	ALD HfO_2 (15 nm-thick)	700/800°C, N_2	[36]
TiO_2	ALD Al_2O_3 (4 nm-thick)	ALD Ti-Al-O (120 cycles)	ALD Al_2O_3 (20 nm-thick)	700/800/900°C, N_2	[37]
$TiAl_2O_5$	ALD Al_2O_3 (4 nm-thick)	PLD $Ti_{0.2}Al_{0.8}O_x$ (8 nm-thick)	PLD Al_2O_3 (8 nm-thick)	900°C, N_2	[38]
ZrO_2	ALD Al_2O_3 (3 nm-thick)	PLD Zr-Si-O (7 nm-thick)	ALD Al_2O_3 (8 nm-thick)	800°C, O_2	[39]

Table 6. ALD processing parameters of Al_2O_3, HfO_2, Hf-La-O and Ti-Al-O

Material	Source and work T	Deposition T	ALD cycle
Al_2O_3	TMA R. T. H_2O R. T	250/300°C	0.1s TMA /4s N_2 /0.1s H_2O /4s N_2
HfO_2	Hf(NEtMe)$_4$* 120°C H_2O R. T	250°C	0.1s Hf(NEtMe)$_4$ /4s N_2/0.1s H_2O /4s N_2
Hf-La-O	Hf(NEtMe)$_4$ 120°C La[N(SiMe$_3$)$_2$]$_4$ 173°C H_2O R. T	250°C	0.1s metal pulse/4s N_2 /0.1s H_2O/ 4s N_2 Sequence: 1 cycle La_2O_3 + 4 cycle HfO_2
Ti-Al-O	TMA R. T. Ti(O-i-Pr)$_4$* 65°C H_2O R. T	250°C	0.1s metal pulse/4s N_2 /0.1s H_2O/4s N_2 Sequence: 24 cycle TiO_2 + 16 cycle Al_2O_3

* Et: ethyl; i-Pr: isopropyl.

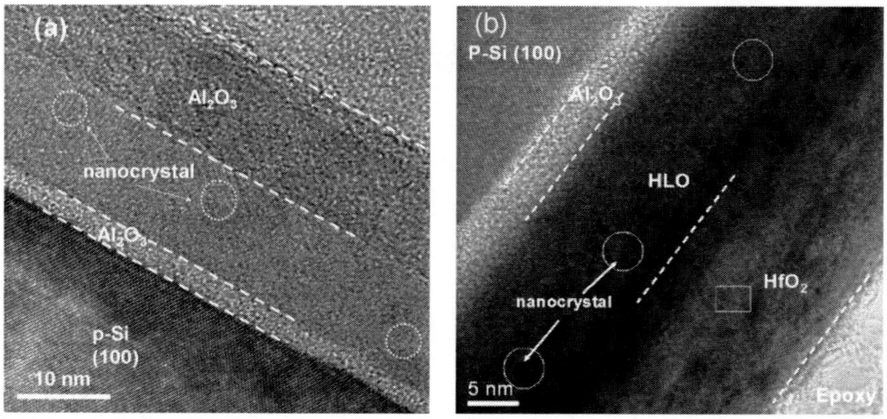

Reprinted with permission from [37]; copyright 2014, Elsevier.

Figure 11. Cross-sectional HRTEM images of ALD all-high-*k* based NC memory cells of (a) Si/Al_2O_3/Ti-Al-O/Al_2O_3 and (b) Si/Al_2O_3/Ha-La-O /HfO_2.

Table 5 lists the architecture and growth processing of 4 kinds of high-*k* NC memory cells derived from ALD and ALD/pulse laser deposition (PLD) [36-39]. The post-annealing is necessary to form high-*k* NCs in storage layer. The p-Si (100) with resistivity of 1-20 Ω·cm was used as the starting substrate. The gate electrode for high-*k* NC memory cells was sputtered-Pt. Table 6 gives the thermal ALD processing parameters of Al_2O_3, HfO_2, Hf-La-O and Ti-Al-O. ALD was performed in a commercial Picosun SUNALE^{TM} R-150B reactor using N_2 (5N) as the carrier/purge gas.

Figure 11(a) and (b) shows the cross-sectional HRTEM images of ALD all-high-*k* based NC memory cells of Si/Al_2O_3/Ti-Al-O/Al_2O_3 and Si/Al_2O_3/Ha-La-O /HfO_2, respectively. In Figure 11(a), after annealed at 800 °C for 3 mins, the interfaces between p-Si and Al_2O_3, Al_2O_3 and Ti-Al-O, and Ti-Al-O and Al_2O_3 are distinct and sharp with the thickness of the Si/Al_2O_3/Ti-Al-O/Al_2O_3 stack of 2.9 nm/10.4 nm/11.7 nm [37]. Spherical TiO_2 NCs are observed in the charge storage layer of Ti-Al-O films with the average size of 3nm. Meanwhile, Al_2O_3 films as the tunneling and blocking oxide layer still keep amorphous features. In Figure 11(b), the interface between p-Si and Al_2O_3, Al_2O_3 and Ha-La-O are also clear with the thickness of the Si/Al_2O_3/Ha-La-O/HfO_2 stack of 3 nm/14.2 nm/16.3 nm, in good agreement with the experiment cycles [36]. After annealed at 800°C for 1 min, LaO NCs are formed in the charge storage layer, as confirmed by grazing incident x-ray diffraction (GIXRD) [36]. The average size of NCs is about 3nm. Meanwhile, HfO_2 films as the blocking layer are well crystallized.

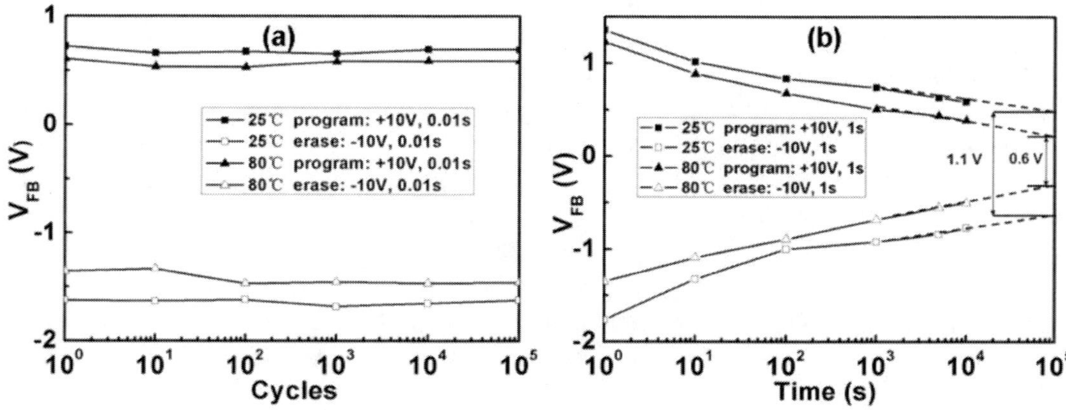

Reprinted with permission from [37]; copyright 2014, Elsevier.

Figure 12. Endurance (a) and retention (b) characteristic of the 800°C $Si/Al_2O_3/Ti\text{-}Al\text{-}O/Al_2O_3/Pt$ stack measured at R. T. and 80 °C.

For memory cells of Si /Al_2O_3/Ti-Al-O/Al_2O_3/Pt [37], compared to the 700°C and 900 °C samples of 0.7 V and 2.8 V, the 800°C sample exhibits a larger memory window of 4.8 V at ± 6 V sweeping. The lacking of NCs in Ti-Al-O layer of 700 °C sample leads to the smallest memory window. The severe interface diffusion and larger NC size (10 nm) in 900°C sample degrades the storage performance. The memory cell annealed at 800°C has a higher P/E speed and larger charge storage density of about 1.63×10^{13} /cm^2 at ± 6 V sweeping. The excellent endurance characteristic of 800°C sample is observed at R. T. and 80°C after 10^5 cycles of P/E pulse (Figure 12(a)), however further improvement is needed for retention characteristics (Figure 12(b)). The tunneling oxide thickness of Al_2O_3, the crystalline quality and density of NCs in Ti-Al-O films need elaborate tuning and optimizing.

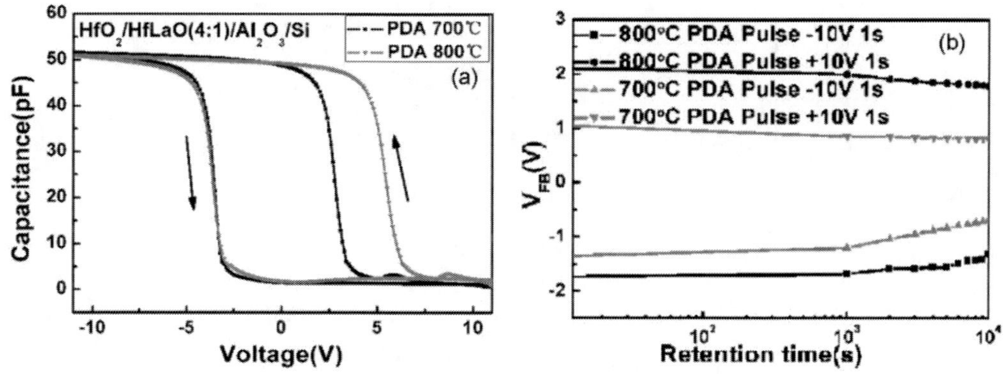

Reprinted with permission from [36]; copyright 2011, IEEE.

Figure 13. C-V curves (a) and retention characteristics (b) of $Si/Al_2O_3/Hf\text{-}La\text{-}O/HfO_2/Pt$ stack structure annealed at 700°C, 800°C in N_2 for 60s at R. T.

For memory cells of Si/Al_2O_3/Hf-La-O/HfO_2/Pt [36], compared with 700°C annealed samples, the 800°C annealed samples have larger memory window of 9V under ±11V sweeping and relatively better retention characteristics with 10-12% charge loss after 10^4s (Figure 13). It can be ascribed to more body defect traps and fewer interface traps in Si/Al_2O_3/Hf-La-O/HfO_2 after higher temperature annealing. The charge density of NCs for 800°C annealed samples is about $2.5\times10^{13}/cm^2$.

A high-*k* NC memory device using $Ti_{0.2}Al_{0.8}O_x$ film as charge trapping layer and amorphous Al_2O_3 as the tunneling and blocking layers was fabricated by combination PLD and ALD [38]. $TiAl_2O_5$ nanocrystals with average size of 3.5 nm are precipitated from the phase separation of PLD $Ti_{0.2}Al_{0.8}O_x$ film annealed at 900 °C (Figure 14(a)). A memory window of 2.3 V and a stored electron density of $1\times10^{13}/cm^2$ are obtained. Good retention characteristics of the memory device at 80 °C with 22.2% V_{fb} shift after 3×10^8 s (~10 years) are observed due to the deep charge trapping level based on the band diagram as identified by the valence band offsets and electron energy loss spectrum measurements (Figure 14(b) and (c)).

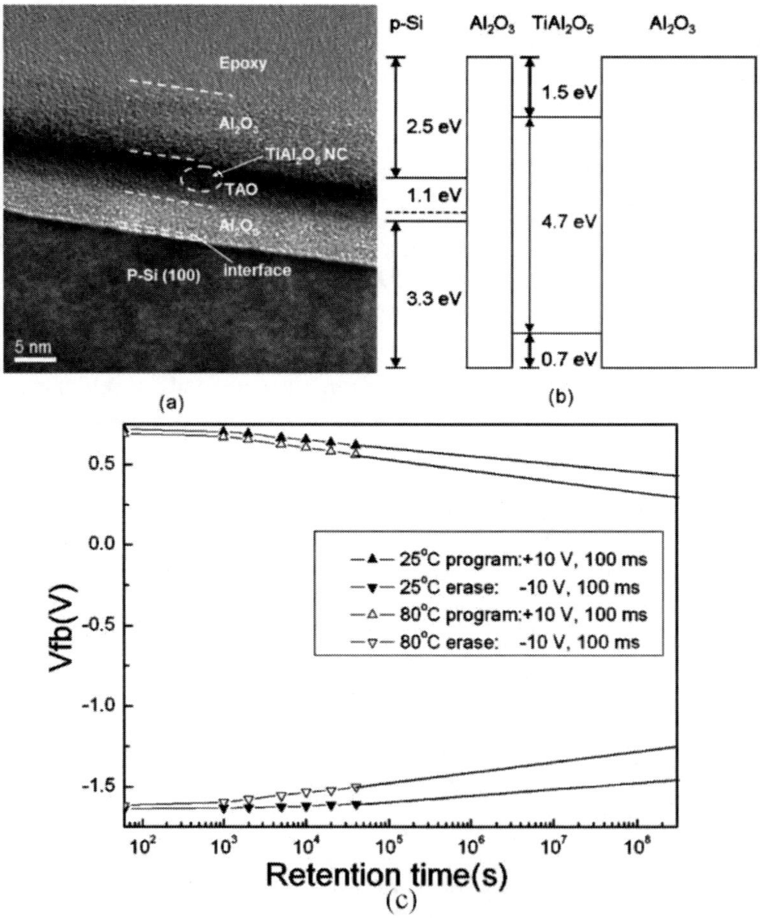

Figure 14. HRTEM image (a) schematic band diagram, (b) and data retention property (c) of the memory device of $Si/Al_2O_3/Ti_{0.2}Al_{0.8}O_x/Al_2O_3$.

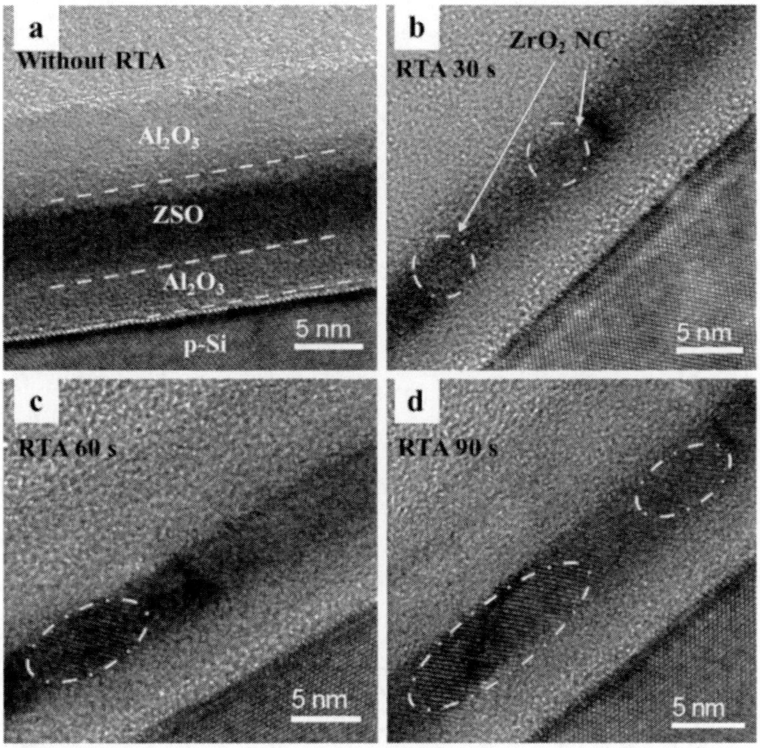

Reprinted with permission from [39]; copyright 2012, Springer.

Figure 15. Cross-sectional HRTEM images of the ZrO_2 NC-based memory cells: (a) before RTA, (b) RTA 30 s, (c) RTA 60 s, and (d) RTA 90 s.

In addition, a high-*k* NC memory device using $(ZrO_2)_{0.6}(SiO_2)_{0.4}$ film as charge trapping layer and amorphous Al_2O_3 as the tunneling and blocking layers was also fabricated by combination PLD and ALD [39]. The effect of RTA post-annealing time at 800 °C on the ZrO_2 NCs formation in $(ZrO_2)_{0.6}(SiO_2)_{0.4}$ films was investigated by TEM (Figure 15). The density and size of ZrO_2 NCs are the critical factors for controlling the charge storage characteristics. The ZrO_2 NCs based memory cells after post-annealing at 800°C for 60 s has average 8 nm NCs size with a stored electron density of $1\times10^{13}/cm^2$, exhibiting the optimalt electrical characteristics of a memory window of 4.4 V under ±5V sweeping and a low charge loss ~5% after 10^5 write/erase cycles operation (Figure 16).

New SONOS Type CTM with High-*k* Composite CTL

Besides the as-above described NC-based CTM, silicon-oxide-nitride-oxide-silicon (SONOS) type CTM has been also regarded as a promising candidate to replace conventional floating-gate devices [17]. However, some performance and reliability issues such as low charge trapping efficiency and poor retention characteristics still exist in SONOS type CTM devices with the continued scaling. As a result, high-*k* materials have been used in SONOS type CTM device as the trapping layer and tunneling/blocking layers to achieve a better storage performance and retention characteristics [40-42].

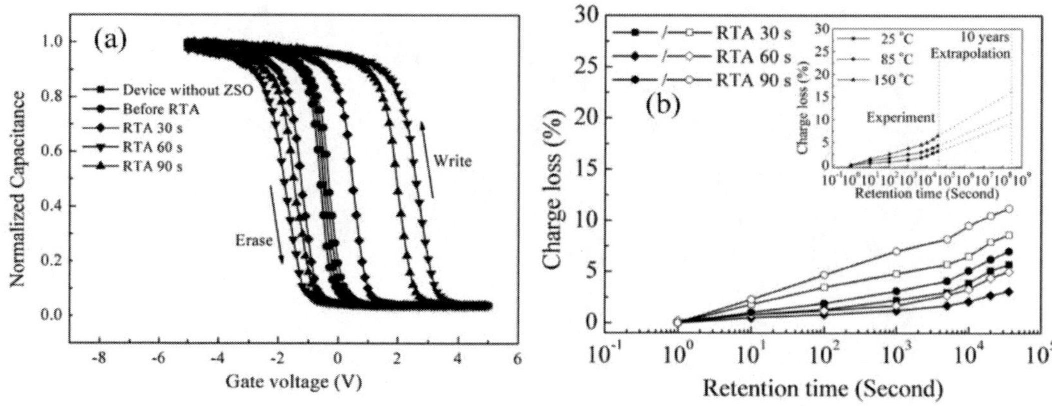

Reprinted with permission from [39]; copyright 2012, Springer.

Figure 16. C-V curves (a) and retention characteristics (b) of ZrO_2 NC-based memory cells postannealed at 800°C for various RTA time.

Table 7. Architecture and growth processing of 4 kinds of new SONOS type CTM memory cells with high-*k* composite dielectrics as CTL

New SONOS type CTM	Tunneling layer	Charge trapping layer	Block layer	RTA Post-annealing	Ref.
ZrO_2/Al_2O_3	ALD Al_2O_3 (3 nm-thick)	ALD ZrO_2/Al_2O_3 nanolaminate (10 nm-thick)	ALD Al_2O_3 (12 nm-thick)	900°C, N_2	[43]
HfO_2/Al_2O_3	ALD Al_2O_3 (3 nm-thick)	ALD HfO_2/Al_2O_3 nanolaminate (10 nm-thick)	ALD Al_2O_3 (15 nm-thick)	800°C, O_2	[44]
$(TiO_2)_{0.8}$ $(Al_2O_3)_{0.1}$	Dry oxidized SiO_2 (3 nm-thick)	Sputtered $(TiO_2)_{0.8}(Al_2O_3)_{0.1}$ (1 nm-thick)	ALD Al_2O_3 (11 nm-thick)	200°C, N_2	[45]
$(Ta_2O_5)_x$ $(Al_2O_3)_{1-x}$	ALD Al_2O_3 (3 nm-thick)	Sputtered $(Ta_2O_5)_x(Al_2O_3)_{1-x}$ (6 nm-thick)	ALD Al_2O_3 (14 nm-thick)	200°C, N_2	[46]

Table 8. ALD processing parameters of Al_2O_3, HfO_2, ZrO_2/Al_2O_3 and HfO_2/Al_2O_3

Material	Source and work T	Deposition T	ALD cycle
Al_2O_3	TMA R. T. H_2O R. T	200/250	0.1s TMA /4s N_2 /0.1s H_2O /4s N_2
HfO_2	$Hf(NMe_2)_4$ 78°C H_2O R. T	250°C	0.1s $Hf(NMe_2)_4$ /4s N_2/0.1s H_2O /4s N_2
Material	**Source and work T**	**Deposition T**	**ALD cycle**
ZrO_2/Al_2O_3	TMA R. T. $Zr(Cl)_4$ 180°C H_2O R. T	300°C	0.1s metal pulse/4s N_2 /0.1s H_2O/ 4s N_2 Various stack structure
HfO_2/Al_2O_3	TMA R. T. $Hf(NMe_2)_4$ 78°C H_2O R. T	250°C	0.1s metal pulse/4s N_2 /0.1s H_2O/ 4s N_2 Various stack structure

Table 7 lists the architecture and growth processing of 4 kinds of CTM devices using composite high-k dielectrics as CTL derived from ALD and ALD/sputtering methods [43-46]. The p-Si (100) with resistivity of 1-20 Ω·cm was used as the starting substrate. The gate electrode for high-*k* NC memory cells was sputtered-Pt. Table 8 gives the thermal ALD processing parameters of Al_2O_3, HfO_2, ZrO_2/Al_2O_3 and HfO_2/Al_2O_3 nanolaminates. ALD was performed in a commercial Picosun SUNALE^{TM} R-150B reactor using N_2 (5N) as the carrier/purge gas.

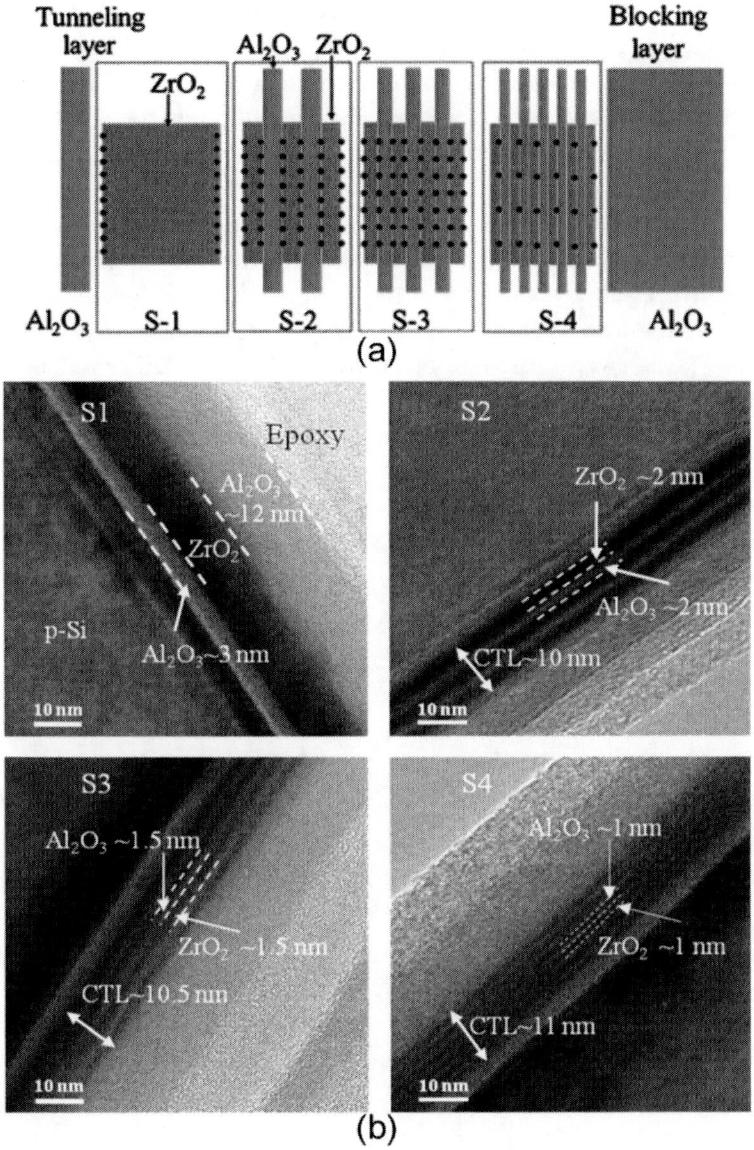

Reprinted with permission from [43]; copyright 2013, Elsevier.

Figure 17. Schematic illustration (a) and cross sectional HRTEM images (b) of ZrO_2/Al_2O_3 nanolaminate-based CTM cells with various CTL structures. S1: a single ZrO_2 (~ 10 nm) film as charge trapping layer; S2: ZrO_2 (2 nm)/Al_2O_3 (2 nm); S3: ZrO_2 (1.5 nm)/Al_2O_3 (1.5 nm); and S4: ZrO_2 (1 nm)/Al_2O_3 (1 nm).

ALD has become a powerful method for the growth of nanolaminate structures with simple and precise control in composition and thickness with atomic scale dimensions. Four ZrO_2/Al_2O_3 nanolaminate-based CTM memory cells with Al_2O_3 tunneling and blocking layers have been prepared by ALD [43], as illustrated in Figure 17(a). The cross sectional HRTEM images show that the thickness of the charge trapping layers (CTL) in the samples S1, S2, S3, and S4 is about 10 nm, 10nm, 10.5nm and 11nm, respectively (Figure 17(b)). Meanwhile, after 900°C RTA treatment the ZrO_2/Al_2O_3 nanolaminate became more obscure as the lamella thickness decreased, which can be ascribed to the interfacial diffusion between the ZrO_2/Al_2O_3 lamella. The interfaces between ZrO_2/Al_2O_3 nanolaminate play an important role in the charge storage characteristics. Keeping the total thickness of the CTL as a constant, decreasing the laminate thickness by increasing the interface number, the memory window increases firstly due to enhanced charge trapped in interface and then decreases due to the electrostatic repulsion between the trapped charges (Figure 18 (a)). The optimal memory window with 6.3V at ±14 V sweeping is obtained for S3. The charge loss after 10 years of the S3 sample is only 7.5% by extrapolating the charge loss curve measured experimentally (Figure 18 (b)). The appropriate laminate thickness and deep quantum wells at the ZrO_2/Al_2O_3 interfaces are responsible for the good retention performance.

The CTM devices with different HfO_2/Al_2O_3 nanolaminate-based CTL and Al_2O_3 tunneling/blocking layers were also fabricated by ALD [44]. The cross sectional HRTEM images of HfO_2/Al_2O_3 nanolaminate-based CTM cells with various CTL structures (1L, 3L and 5L) before and after 800°C RTA indicate that the defects in bulk HfO_2 decrease, but more defect sites are produced at interface of HfO_2/Al_2O_3 due to the inter-diffusion caused by RTA, as seen in Figure 19. It is found that increasing the number of Al_2O_3 intercalation layers can improve the charge-storage capability after RTA treatment. The variation trend of the memory window with the increase of the sweeping gate voltage (Figure 20 (a)) is explained by the band alignments with different applied voltages. The better charge retention performance obtained in the device with multilayered CTLs should be ascribed to the effect of Al_2O_3 intercalation layers, which could suppress the tunneling of the trapped charges back to Si substrate (Figure 20 (b)).

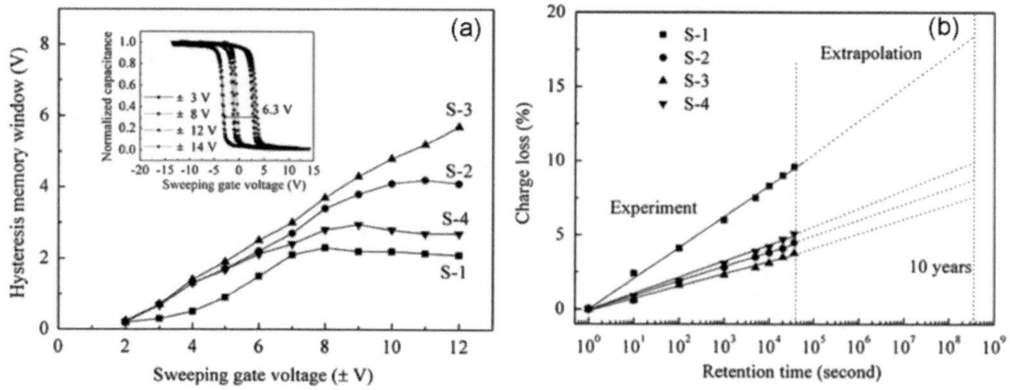

Figure 18. Memory windows (a) and retention characteristics (b) of ZrO_2/Al_2O_3 nanolaminate-based CTM cells with different CTL structures. The inset is the high frequency (1MHz) C–V curves corresponding to different gates weeping voltages.

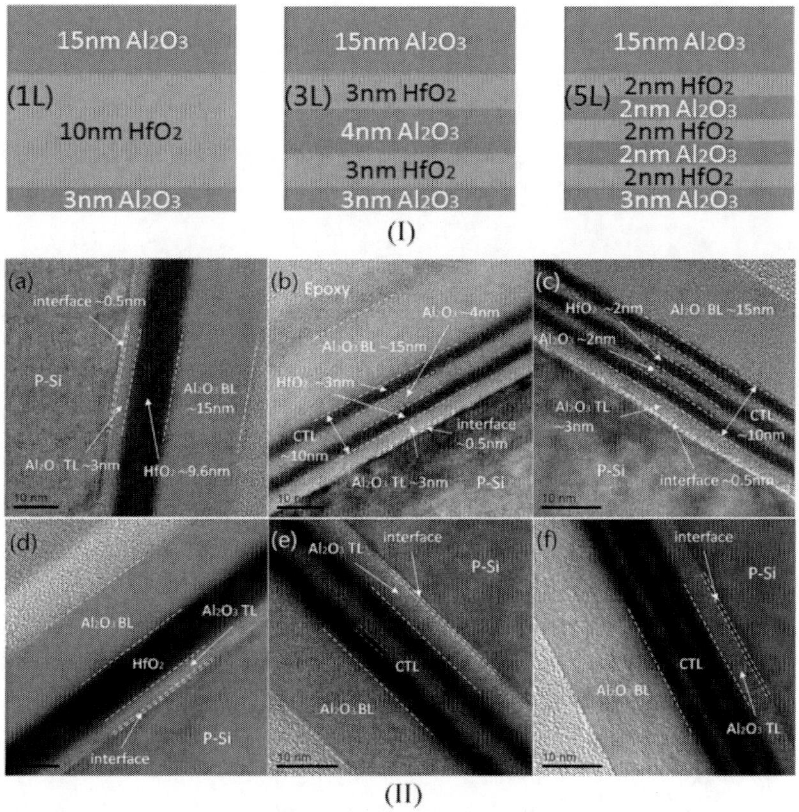

Figure 19. Schematic illustrations (I) and cross sectional HRTEM images (II) of HfO_2/Al_2O_3 nanolaminate-based CTM cells with various CTL structures. As-deposited samples for (a) 1L, (b) 3L, and (c) 5L and RTA samples for (d) 1L, (e) 3L, and (f) 5L. TL, CTL, and BL stand for the tunneling layer, charge trapping layer, and blocking layer, respectively.

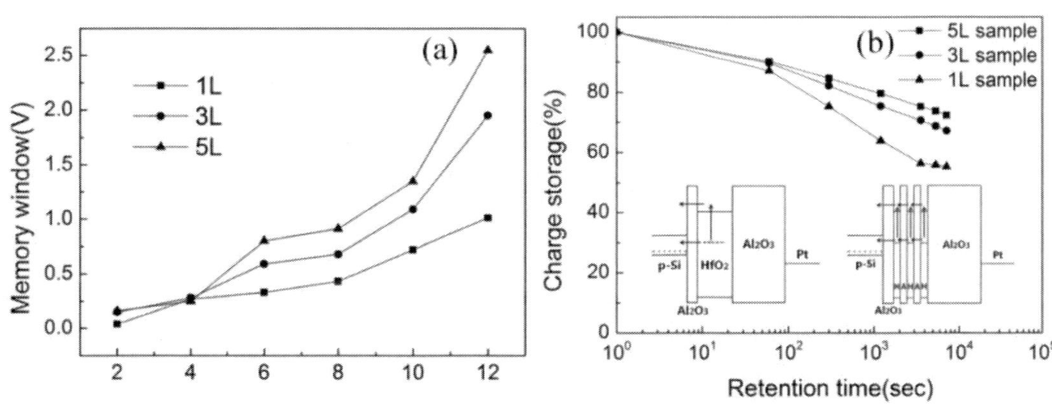

Figure 20. Memory windows (a) and retention characteristics (b) of HfO_2/Al_2O_3 nanolaminate-based CTM cells with different CTL structures. The insets are the band alignments of 1L and 5L samples illustrating the detrapping process of the charges.

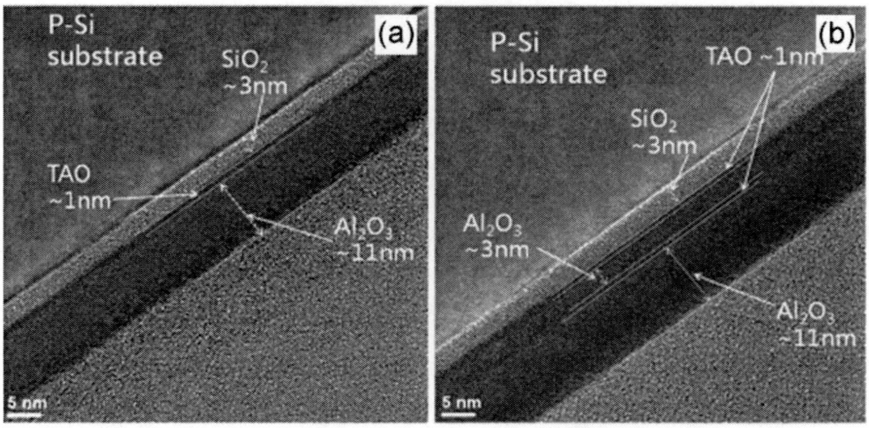

Figure 21. Cross-sectional HRTEM images of TAO-81 memory devices (a) with single TAO-81 CTL and (b) with double TAO-81CTL.

In addition, a CTM device of p-Si/SiO_2/$(TiO_2)_{0.8}(Al_2O_3)_{0.1}$/$Al_2O_3$/Pt was fabricated by combination sputtering and ALD [45], in which a composite of two high-k dielectrics of $(TiO_2)_{0.8}(Al_2O_3)_{0.1}$ (TAO-81) with a thickness of 1 nm was used as the CTL to enhance the charge-trapping efficiency of the memory device, as shown in Figure 21. At an applied gate voltage of ±9V, TAO-81 memory device with single TAO-81 CTL exhibits a surprising memory window of 8.83V with a higher trapped charge density of $9.3\times10^{13}/cm^2$, larger than those in metal NC memory devices [22, 23]. It also shows a fast response to a short voltage pulse of 10^{-5} s (Figure 22(a)). The charge-trapping capability, the endurance, and retention characteristics of TAO-81 memory device can be improved by introducing double TAO-81 charge-trapping layers intercalated by an Al_2O_3 layer (Figure 22(b)). The charge-trapping mechanism in the memory device is mainly ascribed to the generation of the electron-occupied defect level in the band gap of Al_2O_3 induced by the inter-diffusion between TiO_2 and Al_2O_3.

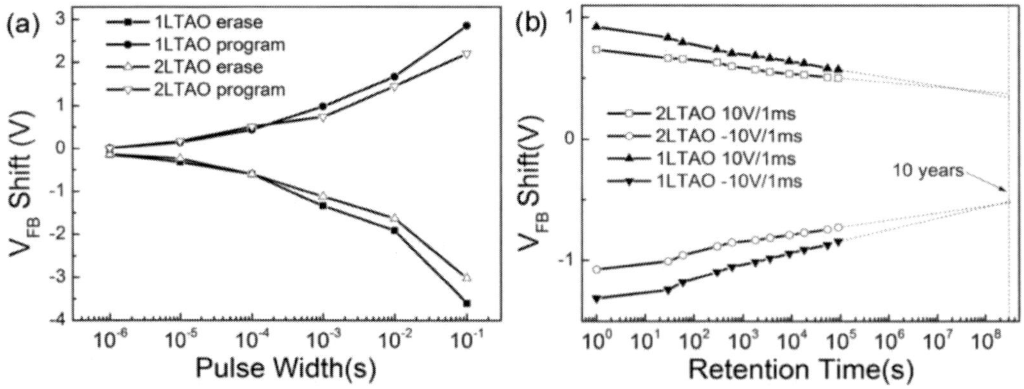

Figure 22. The P/E (a) and retention characteristics (b) of TAO-81 memory devices at room temperature.

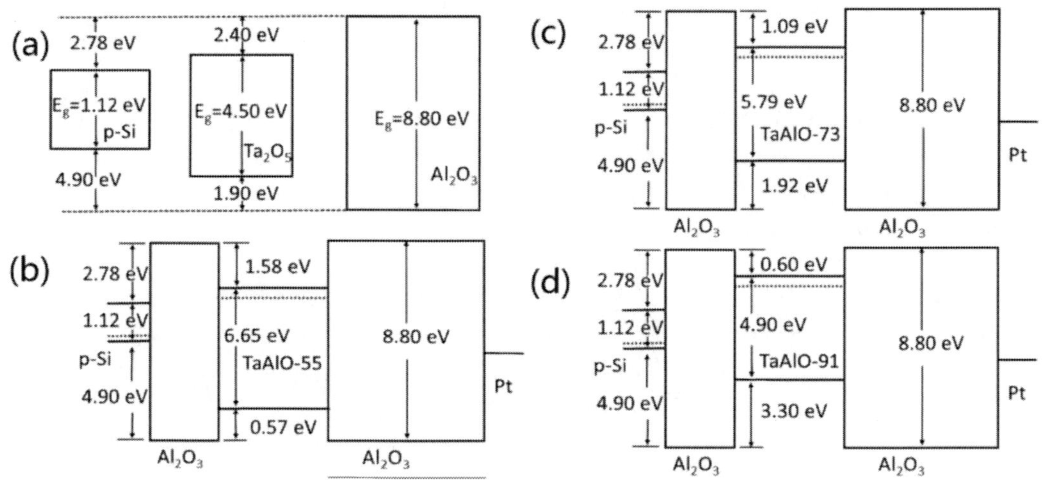

Figure 23. Band alignments of (a) Si, Ta_2O_5, and Al_2O_3, and three CTM devices (b) D1, (c) D2, and (d) D3 without an external electric field. D1: $(Ta_2O_5)_{0.5}(Al_2O_3)_{0.5}$; D2: $(Ta_2O_5)_{0.3}(Al_2O_3)_{0.7}$; D3: $(Ta_2O_5)_{0.1}(Al_2O_3)_{0.9}$.

The CTM devices with the structure p-Si/Al_2O_3/$(Ta_2O_5)_x(Al_2O_3)_{1-x}$/$Al_2O_3$/Pt (x= 0.1 (D3), 0.3 (D2), and 0.5 (D1)) were also fabricated by ALD and RF magnetron sputtering techniques, respectively [46]. A memory window of 7.39V with a charge storage density of $1.97\times10^{13}/cm^2$ at a gate voltage of ±11V Is obtained for the D1 memory device with the composite CTL of $(Ta_2O_5)_{0.5}(Al_2O_3)_{0.5}$. The differences in CTM devices of D1, D2 and D3 with different compositions is ascribed to the relative individual band alignments of the composite $(Ta_2O_5)_x(Al_2O_3)_{1-x}$ with Si. The defect states trapping electrons in $(Ta_2O_5)_x(Al_2O_3)_x$ memory devices should lie near the bottom of the amorphous high-*k* dielectric $(Ta_2O_5)_x(Al_2O_3)_x$, as shown in Figure 23(b)–3(d) by the dotted lines. With a largest mole ratio of Ta_2O_5 in D1, a lowest height of the charge barrier is formed between $(Ta_2O_5)_{0.5}(Al_2O_3)_{0.5}$ and p-Si, and a largest height of the charge barrier between $(Ta_2O_5)_{0.1}(Al_2O_3)_{0.9}$ and p-Si will be expected in D3 with the lowest mole ratio of Ta_2O_5. The lower potential difference between the conduction-band bottoms of Si and $(Ta_2O_5)_x(Al_2O_3)_x$ high-k composite (D1) favors the P/E speed and endurance/retention characteristics of the CTM devices. CTM devices with $(Ta_2O_5)_{0.5}(Al_2O_3)_{0.5}$ show fastest P/E speed and excellent retention property, as indicated Figure 24.

RESISTIVE RANDOM ACCESS MEMORY

Resistive random access memory (RRAM) is one of the most promising candidates for next-generation non-volatile memory applications due to its simple structure, low power consumption, high-speed operation, nondestructive readout, and high-density integration. Many semiconducting and insulating materials including binary transition metal oxides, perovskite oxides, chalcogenides, sulfides, amorphous silicon, organic materials, and ferroelectric materials have been investigated extensively for RRAM applications, especially metal oxides [47]. However, devices using metal oxides suffer from the dispersion of resistive switching parameters, such as the resistance values of low and high resistance states (LRS and HRS), the required set voltages from the HRS to the LRS and reset voltages from the LRS to the HRS, which may lead to false programming and readout error [46]. Ruptures of the conducting filaments with various sizes at random locations are thought as the main reason for the non-uniformity of resistive switching parameters [8, 49]. Several methods have been attempted to solve this problem, such as minimizing grain boundaries, doping, embedding NCs, device area scaling, and bilayer structures in the memory devices [8].

Table 9 presents the architecture and growth processing of 3 kinds of RRAM cells prepared by ALD [48-50]. The p-Si (100) or $Si/SiO_2/Ti/Pt$ was used as the starting substrate. Table 10 gives the thermal ALD (TALD) or plasma-enhanced ALD (PEALD) processing parameters of TiN, Al_2O_3, HfO_2, TiO_2, and $CoPt_x$. ALD was performed in a commercial Picosun SUNALE^{TM} R-200 advanced reactor using N_2 (5N) as the carrier/purge gas for TALD growth and using Ar as the plasma carrier gas and N_2 as the purge gas for PEALD growth.

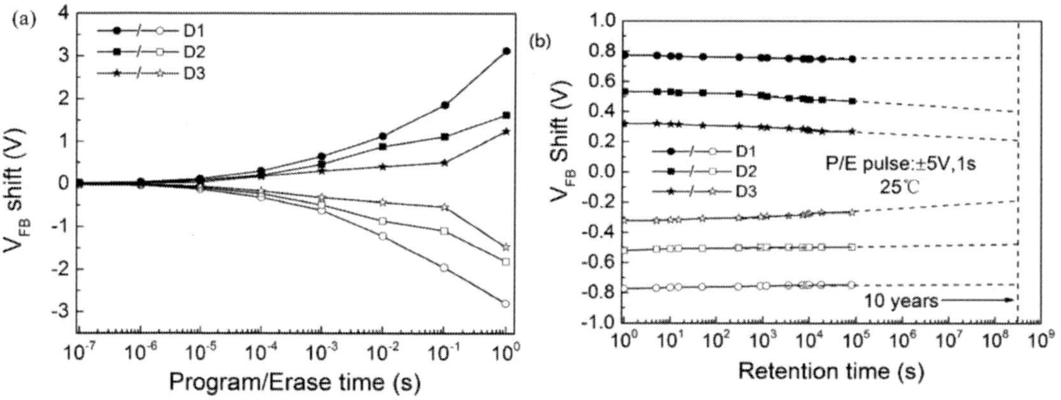

Reprinted with permission from [46]; copyright 2014, American Institute Physics.

Figure 24. The P/E (a) and retention characteristics (b) of $(Ta_2O_5)_x(Al_2O_3)_{1-x}$ CTM devices at room temperature.

Table 9. Architecture and growth processing of 3 kinds of RRAM devices

RRAM cells	Bottom electrode	Resistive layer	Top electrode	RTA Post-annealing	Ref.
$Al_2O_3/HfO_2/Al_2O_3$	PEALD TiN	ALD trilayer $Al_2O_3/HfO_2/Al_2O_3$ (6nm/10 nm/3nm)	Sputtered Pt	600°C, N_2	[50]
$HfO_2/TiO_2/HfO_2$	Pt or PEALD TiN	ALD trilayer $Al_2O_3/HfO_2/Al_2O_3$ (5nm/10 nm/5nm)	Sputtered Pt	No RTA	[51]
$CoPt_x$ NCs/ Al_2O_3	Dry oxidized SiO_2 (3 nm-thick)	ALD $CoPt_x$ NCs/ Al_2O_3	Sputtered Pt	200°C, N_2	[52]

Table 10. ALD processing parameters of TiN, Al_2O_3, HfO_2, TiO_2, and $CoPt_x$

Material	Source and work T	Deposition T	ALD cycle
PEALD TiN	$TiCl_4$ R. T. NH_3 plasma using liquid NH_3 (R.T.) 150 sccm	400°C	0.1s $TiCl_4$ /4s N_2 /24s NH_3 plasma /4s N_2
TALD Al_2O_3	TMA R. T. H_2O R. T	250°C	0.1s TMA /4s N_2 /0.1s H_2O /4s N_2
TALD HfO_2	$Hf(NEtMe)_4$ 150°C H_2O R. T	250°C	0.1s $Hf(NEtMe)_4$ /4s N_2/0.1s H_2O /4s N_2
TALD TiO_2	$TiCl_4$ R. T. H_2O R. T	300°C	0.1s $TiCl_4$ /4s N_{22} /0.1s H_2O/ 4s N_2
PEALD $CoPt_x$	$CoCp(CO)_2$ R. T. $PtMeCpMe_3$ 70°C H_2 plasma using 100 sccm H_2	300°C	Co cycle: 1s $CoCp(CO)_2$/10s N_2 /21.5s H_2 plasma / 10s N_2 Pt cycle: 0.2s $PtMeCpMe_3$/ 6s N_2 /21.5s H_2 plasma / 10s N_2 $CoPt_x$ Sequence: 1 cycle Co + 1 cycle Pt

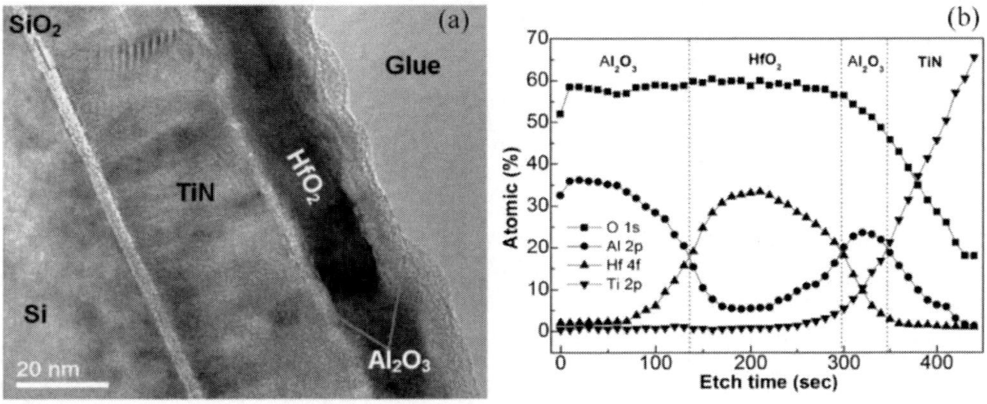

Figure 25. Typical cross-section TEM image and XPS depth profile of the device unit of $Al_2O_3/HfO_2/Al_2O_3$ trilayer structure on TiN-coated Si by ALD.

A flexible RRAM device with trilayer structure of $Al_2O_3/HfO_2/Al_2O_3$-based functional stacks on TiN-coated Si substrate has been fabricated by PEALD and TALD [50]. The cross-section HRTEM image and XPS depth profile of $Al_2O_3/HfO_2/Al_2O_3$ on TiN-coated Si confirm the existence of interfacial layers between trilayer structures of $Al_2O_3/HfO_2/Al_2O_3$ after 600°C post-annealing, as shown in Figure 25 (a) and (b).

The memory units of $Si/TiN/Al_2O_3/HfO_2/Al_2O_3/Pt$ exhibit a typical bipolar, reliable, and reproducible resistive switching behavior, such as stable resistance ratio (>10) of OFF/ON states, sharp distribution of set and reset voltages (Figure 26). The better switching endurance up to 10^3 cycles and longer data retention at 85°C over ten years have been confirmed in Figure 27.

The possible resistive switching mechanism of trilayer structure of $Al_2O_3/HfO_2/Al_2O_3$ has been proposed, as seen in Figure 28. Based on the XPS results (Figure 29), HfO_2 layer produces and stores more oxygen vacancies than Al_2O_3 layer. The physical size of the conductive filament in oxygen vacancy deficient layer is narrower than in the oxygen vacancy rich layer [53], so it can be inferred reasonably that the filament in HfO_2 layer is thicker and more intensive. Based on the bilayer-structure model, the oxygen exchange through the interface between two layers played an important role in improving resistive switching characteristics [51]. In this work, the two interfacial layers (IL) between trilayer-structure of $Al_2O_3/HfO_2/Al_2O_3$ on TiN-coated have also similar effect on connected or broken filaments in memory units, as indicated Figures 28.

The two interfacial layers between Al_2O_3/IL/HfO_2/IL/Al_2O_3 could affect the growth position, direction, and overlapping of conductive filaments. The linkage (Figure 28(a)) or rupture (Figure 28(b)) of the conductive filaments corresponds to the SET from HRS to LRS or the RESET from LRS to HRS, respectively, which more easily occurs in two interfacial layers. Further, the shape and position of the conductive filaments in Al_2O_3 and HfO_2 layers change less in the SET and RESET processing. The resistive switching mechanism is mainly dominated by formation and rupture of conducting filaments of oxygen vacancies in interfacial layers.

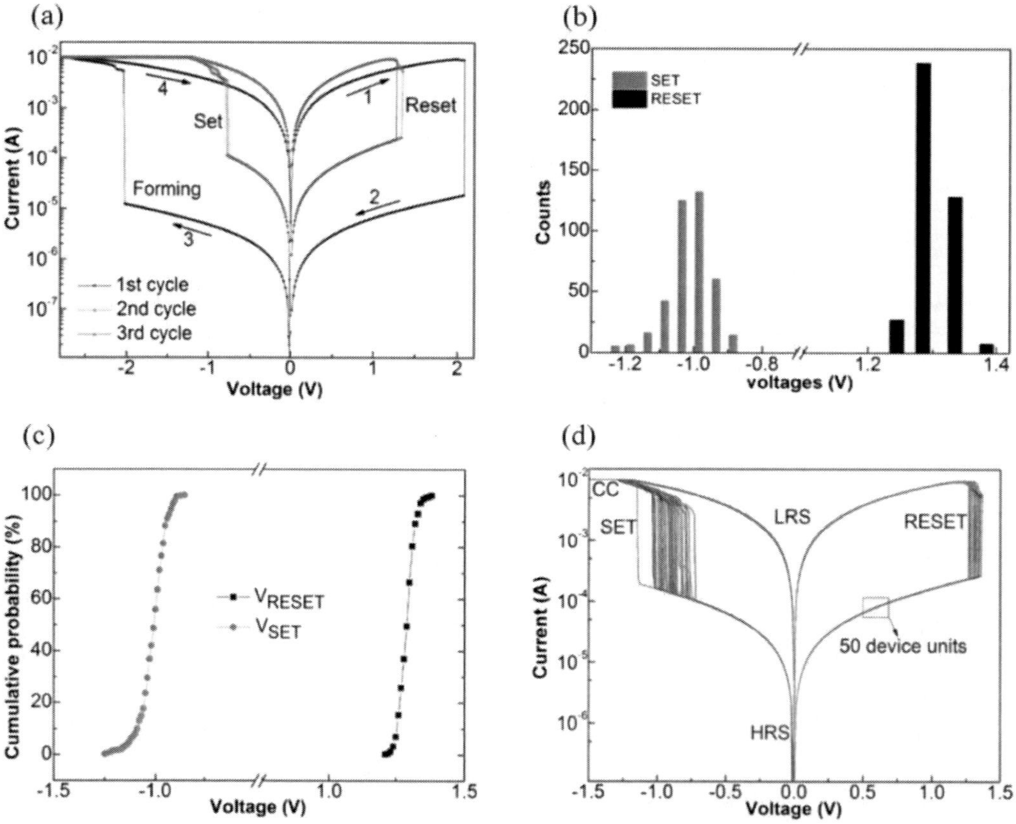

Reprinted with permission from [50]; copyright 2015, Springer.

Figure 26. (a) Typical resistive switching characteristics of the device unit of Si/TiN/$Al_2O_3/HfO_2/Al_2O_3$/Pt after initial, second and third cycles. (b) and (c) Statistical results of distribution and cumulative probability of set and reset voltages from a device unit for 400 times tests. (d) Resistive switching data of 50 randomly selected device units.

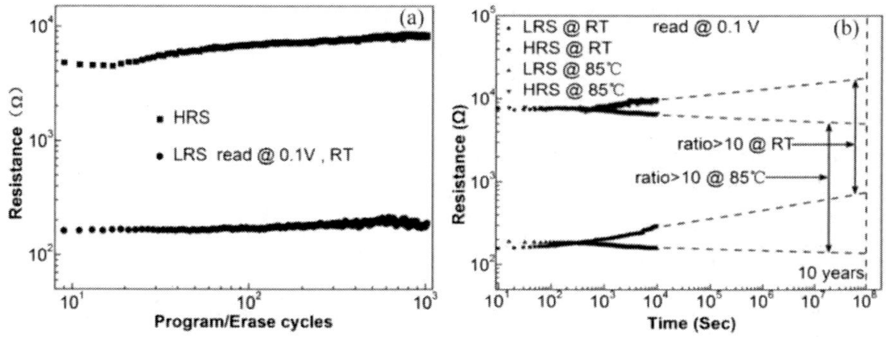

Reprinted with permission from [50]; copyright 2015, Springer.

Figure 27. Durability of the device unit of $Si/TiN/Al_2O_3/HfO_2/Al_2O_3/Pt$. (a) The continuous program and erase test. (b) Read disturbance test for device after 10^4 s retention time at room temperature and 85°C.

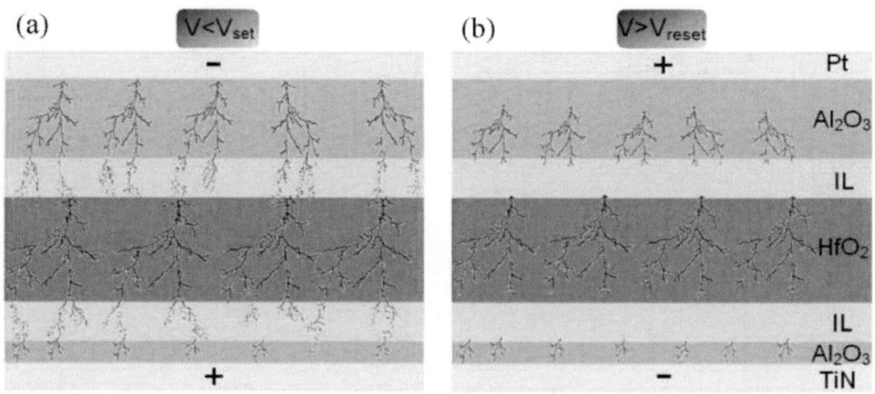

Reprinted with permission from [50]; copyright 2015, Springer.

Figure 28. Schematics of proposed filament-switching mechanism of $Al_2O_3/HfO_2/\ Al_2O_3/Pt$ on TiN-coated Si. (a) Low resistance state (LRS) when applied voltage is less than set voltage. (b) High resistance state (HRS) when applied voltage is greater than reset voltage.

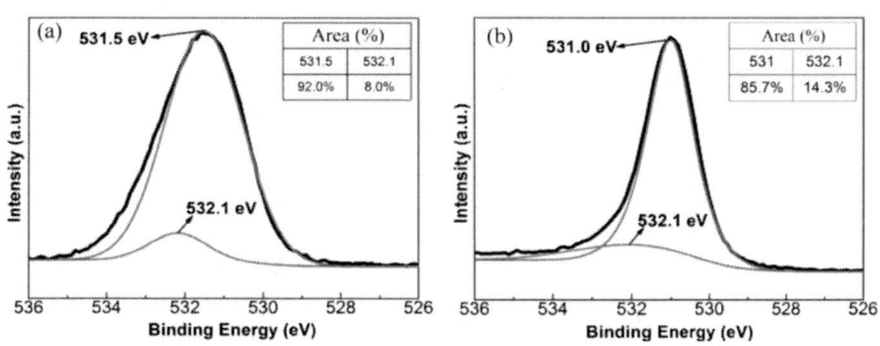

Reprinted with permission from [50]; copyright 2015, Springer.

Figure 29 Narrow-scan O 1s XPS spectra from Al_2O_3 (a) and HfO_2 layers (b) of trilayer structure of $Al_2O_3/HfO_2/Al_2O_3$ on TiN-coated Si.

The $HfO_2/TiO_2/HfO_2$ trilayer-structure RRAM devices have also been fabricated on $Si/SiO_2/Ti/Pt$ and Si/TiN substrates by ALD [51]. The effect of bottom electrodes of Pt and TiN on the resistive switching properties of trilayer structure has been investigated deeply. The memory cells of $Pt/HfO_2/TiO_2/HfO_2/Pt$ and $TiN/HfO_2/TiO_2/HfO_2/Pt$ show bipolar resistive switching characteristics. At the low resistance state, the conductive mechanism follows the Ohmic law. The filament model of oxygen vacancy migration can be used to explain the switching behavior. At the high resistance state, it is dominated by the typical trap-controlled space charge limited conduction (SCLC). Both $Pt/HfO_2/TiO_2/HfO_2/Pt$ and $TiN/HfO_2/TiO_2/HfO_2/Pt$ exhibit good endurance and retention characteristics due to the trilayer structures.

However, the bottom electrodes of Pt and TiN have great influence on the electroforming polarity, ratio of high and low resistance and dispersion of the operating voltage of trilayer-structure memory cells. The RRAM cells with Pt bottom electrodes have a forming voltage of about +7 V, set voltage of +2 V, and reset voltage of -0.8 V with ratio of high and low resistance of $>10^5$, whereas the RRAM ones with TiN bottom electrodes have an opposite forming voltage of about -4 V, set voltage of -1.5 V, and reset voltage of +1.5 V with ratio of high and low resistance of only 10^2 (Figure 30).

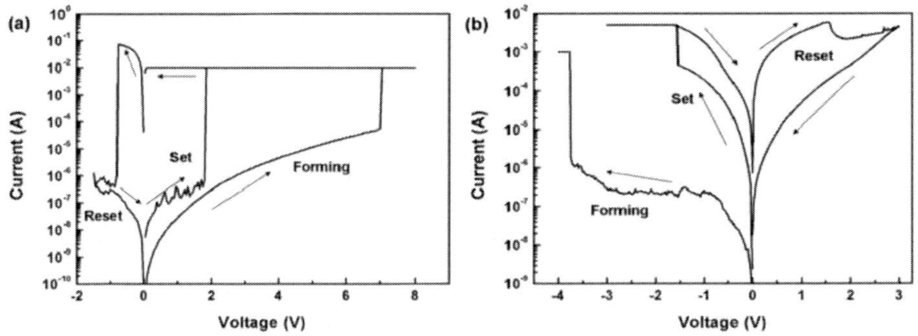

Figure 30. Typical bipolar resistive switching characteristics of the (a) $Pt/HfO_2/TiO_2/ HfO_2/Pt$ and (b) $TiN/HfO_2/TiO_2/HfO_2/Pt$ devices.

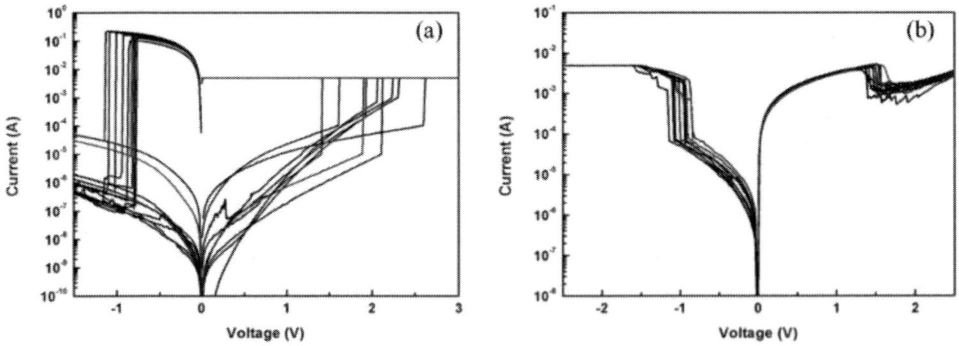

Figure 31. The I-V curves of randomly selected ten device units of the (a) $Pt/HfO_2/TiO_2/HfO_2/Pt$ and (b) $TiN/HfO_2/TiO_2/HfO_2/Pt$ devices.

Compared to the RRAM cells on $Si/SiO_2/Ti/Pt$, different storage units on Si/TiN substrates and single storage unit with TiN bottom electrodes after multiple cycle operations show relatively narrow distribution of the set/reset operation voltages (Figure 31). It is attributed to the fact that the formation of conductive filament is significantly affected by the initial distribution of oxygen vacancy in the insulator. The TiN electrode with high oxygen affinity causes a lot of oxygen vacancies in the HfO_2 film near the TiN bottom electrode [8], easily leading to the formation of conductive oxygen vacancy filaments with improved uniformity of resistive switching parameters.

The trilayer structure device units of $Al_2O_3/HfO_2/Al_2O_3$ or $HfO_2/TiO_2/$ HfO_2 on TiN-coated Si prepared by ALD may be a potential candidate for oxide-based RRAM.

Finally, the Al_2O_3-based RRAM devices with embedded $CoPt_x$ NCs have also been prepared on $Si/SiO_2/TiN$ substrates by ALD [52]. The impact of $CoPt_x$ NCs and their average size/density on the resistive switching properties has been explored. Figure 32 (a) and (b) shows the TEM images of PEALD $CoPt_x$ NCs deposited directly onto TiN coated TEM grids with total ALD cycles of 100 and 200. The average size of density of $CoPt_x$ NCs on TiN is about 3 nm/1.1×10^7/cm^2 for 100 ALD cycles and 9 nm/3.5×10^6/cm^2 for 100 ALD cycles, respectively. Cross-section TEM images of $Si/SiO_2/TiN/9$ nm-$CoPt_x$ NCs/Al_2O_3 RRAM devices indicate that 9 nm-$CoPt_x$ NCs have been successfully embedded into 10 nm-thick Al_2O_3 thin films (Figure 32 (c) and (d)).

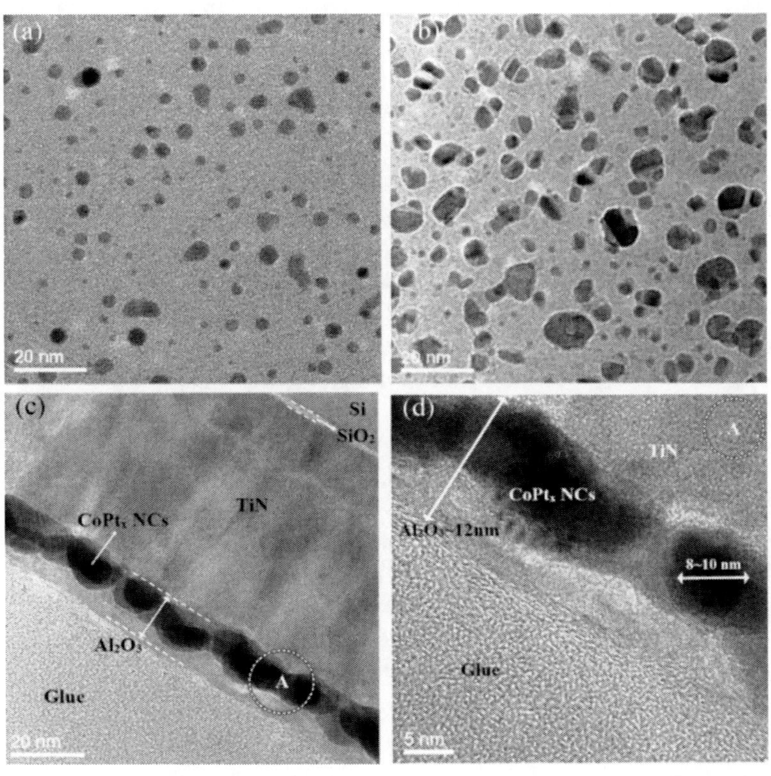

Figure 32. Plan-view TEM images of PEALD $CoPt_x$ NCs deposited directly onto TiN coated TEM grids with total 100 (a) and 200 (b) ALD cycles. Cross-section TEM images of $Si/SiO_2/TiN/9$ nm-$CoPt_x$ NCs/Al_2O_3 RRAM devices by ALD (c) and magnification view of A region (d).

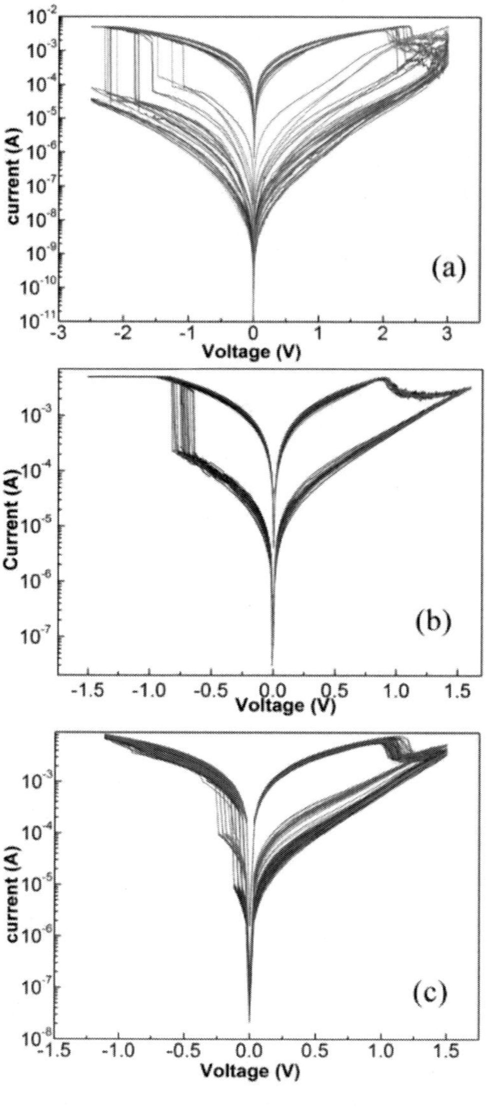

Figure 33. The I-V curves of a single device unit of (a) control sample of TiN/Al_2O_3/Pt, (b) TiN/9 nm-$CoPt_x$ NCs/Al_2O_3/Pt and (c) TiN/3 nm-$CoPt_x$ NCs/Al_2O_3 for 50 times test.

The I-V curves of a single device unit of (a) control sample of TiN/Al_2O_3/Pt, (b) TiN/9 nm-$CoPt_x$ NCs/Al_2O_3/Pt and (c) TiN/3 nm-$CoPt_x$ NCs/Al_2O_3 for 50 times test are plotted in Figure 33(a)-(c), indicating a typical bipolar resistive switching characteristic. Compared to control sample without $CoPt_x$ NCs, the RRAM devices containing $CoPt_x$ NCs show reproducible resistive switching behavior, such as stable resistance ratio (≧ 20) of OFF/ON states, sharp distribution of set and reset voltages. Moreover, the set/reset voltage for control sample without $CoPt_x$ NCs is larger and -2.0 V/2.5 V, respectively. Whereas the set/reset voltage for 3 nm- and 9 nm-$CoPt_x$ NCs RRAM devices becomes much smaller and -0.22 V/1.25 V and -0.75 V/1.0 V, respectively. Therefore, the incorporation of $CoPt_x$ NCs into Al_2O_3 films greatly decreases the set and reset voltages, beneficial to the drop of power consumption.

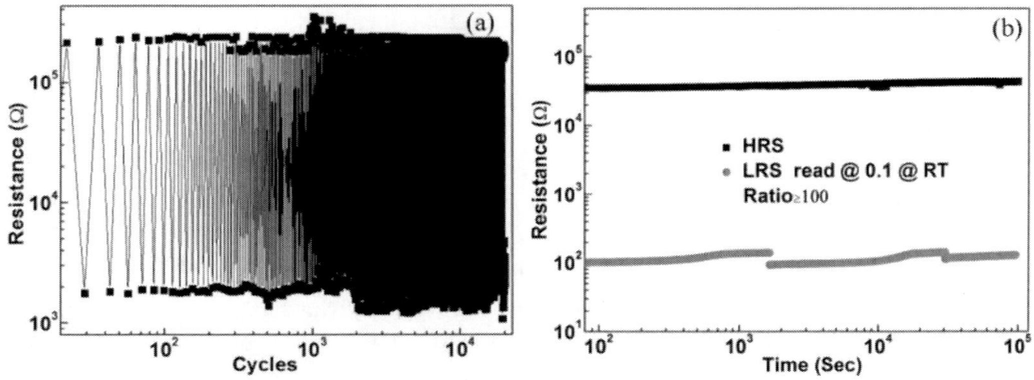

Figure 34. Durability of the device unit of TiN/9 nm-$CoPt_x$ NCs/Al_2O_3/Pt. (a) The continuous program and erase test after 10^4 cycles. (b) Read disturbance test for device after 10^5 s retention time at room temperature.

Compared with 3 nm-$CoPt_x$ NCs RRAM, the memory units of Si/SiO_2/TiN/9 nm-$CoPt_x$ NCs/Al_2O_3/Pt exhibit excellent switching endurance up to 10^4 cycles, and longer data retention (> 10^5 s) at room temperature over ten years with stable resistance ratio ($\geqq$ 20) (Figure 34). The possible switching mechanism of trilayer structure of Al_2O_3/HfO_2/Al_2O_3 has been proposed. The $CoPt_x$ NCs on TiN bottom electrode play an important role in improving resistive switching characteristics. The embedded $CoPt_x$ NCs in Al_2O_3 films produce stronger localized electric field, leading to the formation/rupture of the conducting filaments at $CoPt_x$ NC position with improved data distribution.

FERROELECTRIC RANDOM ACCESS MEMORY

FeRAM are promising candidates for low power and high density future memory technologies [9, 53]. However, integrating ferroelectric thin films such as well-known $Pb(Zr_xTi_{1-x})O_3$ (PZT) and $SrBi_2Ta_2O_9$ (SBT) into conventional Si based-IC proves rather challenging. Buffer layers have to be implemented to protect the oxygen sensitive plug during high temperature metal-ferroelectric-metal (MFM) capacitor fabrication and the ferroelectric layer itself from hydrogen damage during back end of line (BEOL) processing. Conformal deposition technologies and etch processes able to cope with the material properties and larger physical thickness of the commonly used MFM stacks are inadequate yet, therefore, hindering FeRAM future scaling in terms of 3-D integration [54].

Recently, Müller et al. have reported that ALD-derived ultrathin Hf-based oxide thin films with some doping elements such as Si, Al, Zr, and Y exhibit ferroelectric behavior [10, 55-58], which triggers new interest for FeRAM application due to its excellent compatibility with Si technology, because Hf-based high-*k* dielectrics have been integrated in metal-oxide-semiconductor field-effect transistors (MOSFETs) and dynamic random access memory (DRAM) [15, 59].

Table 11 summarizes the processing condition, dopant concentration, and ferroelectric property of ALD-derived ultrathin Hf-based oxide thin films [10, 55-57]. The used ALD precursors contain $Hf(NEtMe)_4$, $Zr(NEtMe)_4$, $Si(NMe_2)_4$, TMA and $Y(MeCp)_3$ with ozone as oxygen source.

Figure 35 (a) and (b) shows the TEM/HRTEM images and typical P-V hysteresis loops of $TiN/Zr_{0.5}Hf_{0.5}O_2/TiN$ capacitors, respectively [55]. $TiN/Hf_{0.5}Zr_{0.5}O_2/TiN$ capacitors display ferroelectric hysteresis for dielectric films as thin as 7.5 nm, yielding a remnant polarization (P_r) of 16 $\mu C/cm^2$ and a coercive field (E_c) of 1 MV/cm. Quasi-static P-V, C-V, and piezoelectric hysteresis curves show the characteristic features expected for MFM capacitors.

The ferroelectricity origin in solid solution of ZrO_2-HfO_2 has been explored [58]. It is found that there exists a composition- and temperature-dependent transition from tetragonal to a stable ferroelectric orthorhombic phase in the HfO_2−ZrO_2 solid solution. The GIXRD structural characterization reveals the orthorhombic phase to be of space group $Pbc2_1$ (Figure 36) [58], whose noncentrosymmetric nature is thought to be responsible for the spontaneous polarization in the nanoscale ferroelectrics. Similar orthorhombic phase has been confirmed in Al-doped HfO_2 ultrathin films [57]. The effects of annealing temperature and film thickness on the crystal structure and ferroelectric properties of $Hf_{0.5}Zr_{0.5}O_2$ films have also been examined [60]. The $Hf_{0.5}Zr_{0.5}O_2$ films consist of tetragonal, orthorhombic, and monoclinic phases. The ferroelectric orthorhombic phase content is almost independent of anneal temperature, but decreases with increasing film thickness. The origin of the ferroelectric polarization switching in orthorhombic HfO_2 with dopant of Si, Al, Zr and Y has been investigated by first principles calculations [61], which can be regarded as being the coordinated displacement of four O ions in the orthorhombic unit cell, leading to a saturated polarization as high as 53 $\mu C/cm^2$.

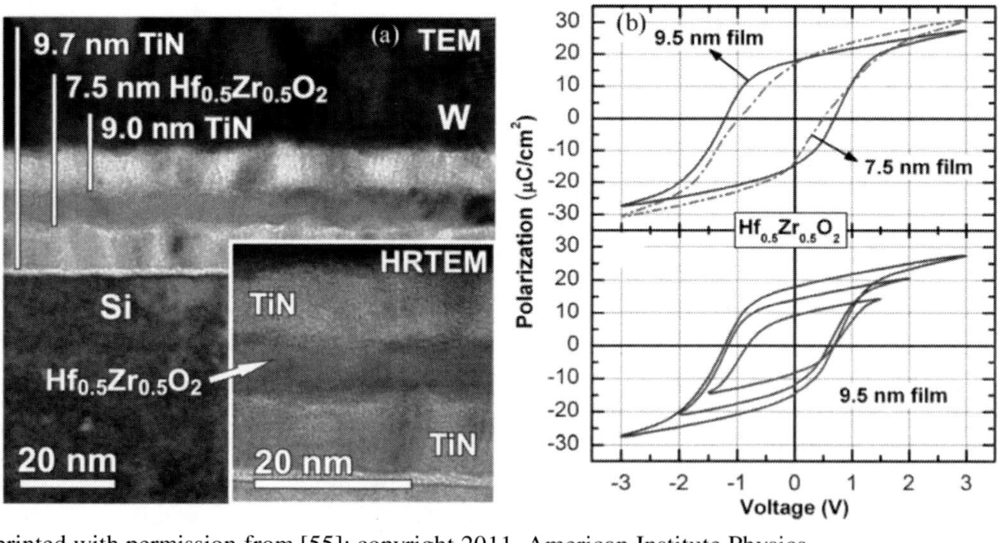

Figure 35. (a) TEM/HRTEM images and (b) typical P-V hysteresis loops of $TiN/Zr_{0.5}Hf_{0.5}O_2/TiN$ capacitors, respectively.

Table 11. Processing, composition and ferroelectric property of ALD-derived ultrathin Hf-based oxide thin films

Ferroelectric thin film	Dopant content* /thickness (nm)	Bottom/top Electrode	Anneal T (°C) /time*	P_r ($\mu C/cm^2$)	E_c (MV/cm)	Ref.
Si: HfO_2	<4% / 7-10	TiN/ TiN	1000, 20s	10	1	[10]
$Zr_{0.5}Hf_{0.5}O_2$	50% / 7.5-9.5	TiN/ TiN	450	16	1	[55]
Y: HfO_2	5.2% / 10	TiN/ TiN or Pt	600, 20s	24	1.2	[56]
Al: HfO_2	4.8% / 16	TiN/ TiN or Pt	600, 20s or 1000, 2s	5	1	[57]

* Dopant content: mol percentage ** anneal in N_2.

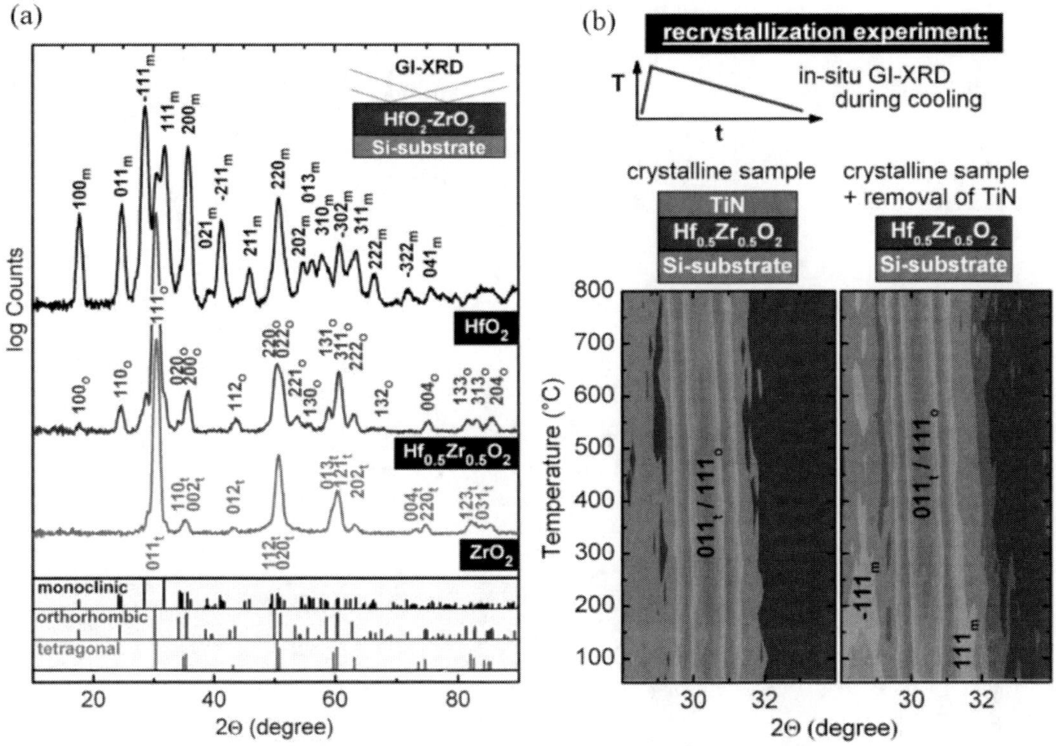

Reprinted with permission from [58]; copyright 2012, American Chemistry Society.

Figure 36. (a) GIXRD diffractograms of 9 nm-thick ZrO_2, $Hf_{0.5}Zr_{0.5}O_2$, and HfO_2 thin films at an incident angle of 0.55°. Starting from ZrO_2 a composition-dependent tetragonal to orthorhombic to monoclinic transition is observed. (b) In situ GIXRD measurements allow the direct observation of a suppressed tetragonal to monoclinic transformation during cooling in samples confined by a TiN electrode.

The reliability characteristics of ferroelectric Hf-based oxide ultrathin films for memory applications such as forming gas annealing (FGA), retention, imprint, and endurance data have been studied deeply [55, 62-64]. The Pt (top electrode)/$Hf_{0.5}Zr_{0.5}O_2$/TiN capacitors show the maximum $2P_r$ loss of ~40% due to the Pt catalyst for H-incorporation after FGA. However, by inserting a ~20-nm-thick TiN layer between Pt and $Hf_{0.5}Zr_{0.5}O_2$, the degradation of 2 P_r is only ~12% (Figure 37). Evidently, $Hf_{0.5}Zr_{0.5}O_2$ ferroelectric thin film is more resistant to degradation by FGA compared with the conventional ferroelectrics such as PZT and SBT [62]. Stable ferroelectric switching behavior for 4.6% Si doped HfO_2 ultrathin films (10 nm-thick) can be observed in a wide temperature range from 80 K up to 470 K [62]. Almost no retention loss for saturated polarization states up to cumulative testing times of 1000 h tested at 125°C. Traditional imprint behavior of the programmed state occurs after a few hours of baking time, and stable behavior could be verified until the end of the 1000-h retention test. Endurance property of the ferroelectric thin films depends significantly on the annealing temperature of the capacitors and on the cycling voltage during testing. A lower annealing temperature of 650°C could improve the breakdown-limited endurance to 10^{10} cycles [63]. Above all, ALD-derived ultrathin Hf-based oxide thin films are promising candidate for FeRAM applications.

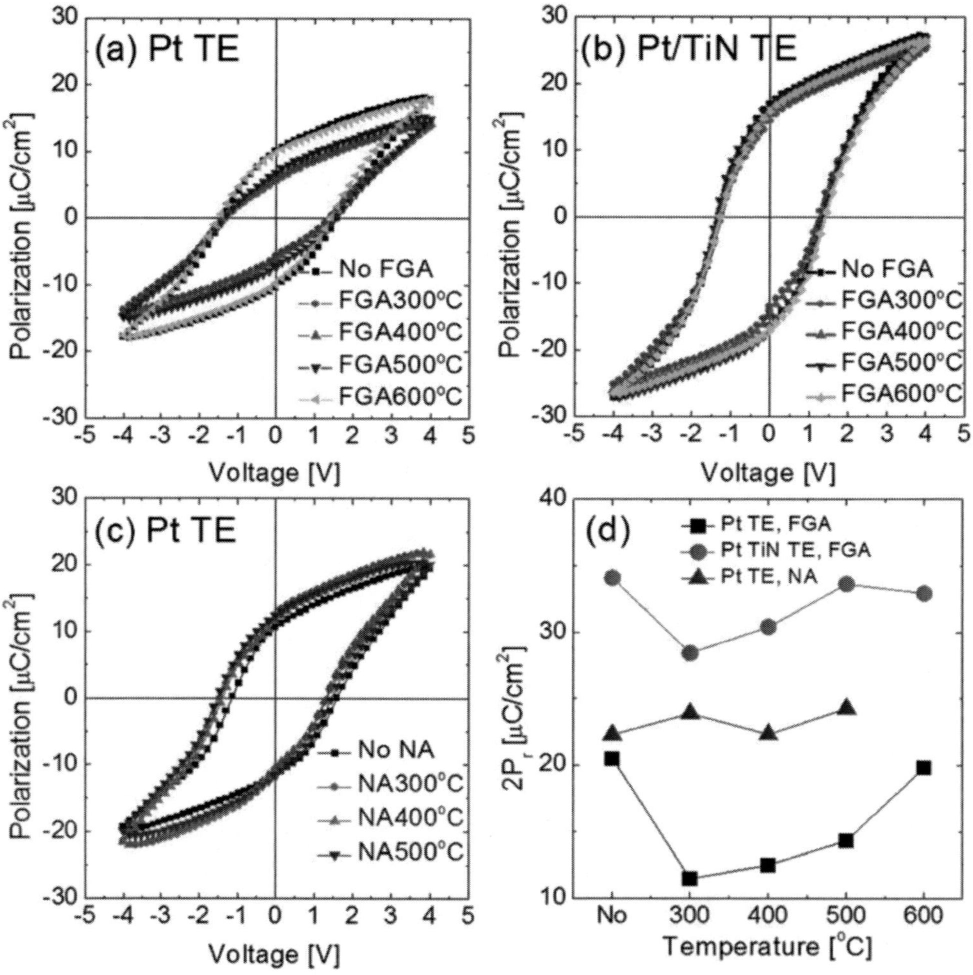

Figure 37. Polarization-voltage characteristics of_~10-nm-thick $Hf_{0.5}Zr_{0.5}O_2$ films after (a) FGA with Pt top electrode, (b) FGA with Pt/TiN top electrode, and (c) NA with Pt top electrode. (d) Summary of the changes in 2Pr as a function of annealing temperature. FGA: forming gas annealing and NA: N2 annealing.

ULTRAHIGH DENSITY MAGNETIC RECORDING MEDIA

Higher areal density is urgently needed in magnetic recording to provide more storage capacity on a hard disk drive. However, when the recording bit is down to 10 nm scale, the recording bit might flip itself due to thermal fluctuation and renders the recording data unstable. There are two methods to acquire higher areal density: one is to acquire materials with larger magnetic anisotropy; the other one is to fabricate patterned media with space among each bit.

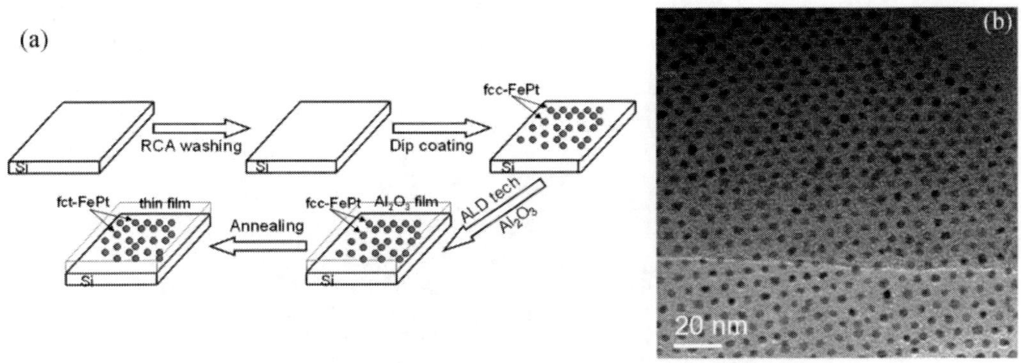

Figure 38. (a) Schematic diagram of fabrication technique for FePt/Al_2O_3 nanocomposite films by combination of ALD technology and FePt self-assembly. (b) TEM image of self-assembled *fcc*-FePt NCs pattern.

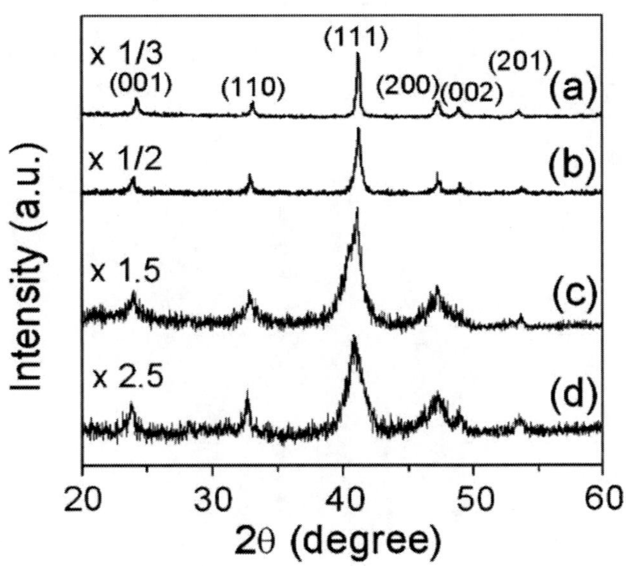

Figure 39. XRD patterns of fct-FePt NCs on Si substrates after annealing at 700 °C for 1 h: (a) without Al_2O_3 protection layer; (b), (c) and (d) with 5 nm-thick, 10 nm-thick and 20 nm-thick Al_2O_3 capping layers, respectively.

In recent years, chemically ordered face-centered tetragonal (fct) $L1_0$ FePt NCs have attracted great interests because of their large magnetocrystalline anisotropy (K_u), which makes it excellent candidate for applications in ultrahigh-density magnetic recording media [65-66]. Typically, as-prepared FePt NCs have a disordered fcc structure with low K_u and coercivity (H_C), and high-temperature annealing (usually above 550°C) is necessary to obtain the chemically ordered high-K_u $L1_0$ phase [67]. However, the heat-treated temperature above 450°C may cause aggregation and sintering, and decrease the particle positional order [68-70].

A new fabrication technique for the FePt/Al_2O_3 nanocomposite films for ultrahigh density magnetic memory applications has been developed by combination of ALD technology and FePt self-assembly firstly [31, 71], as illustrated in Figure 38. The XRD and TEM results show that the ALD amorphous Al_2O_3 capping layer can prevent FePt NCs from sintering and keep the position order of the FePt NCs assembly, producing the $L1_0$-FePt/Al_2O_3 nanocomposite films (Figure 39 and 40) [31].

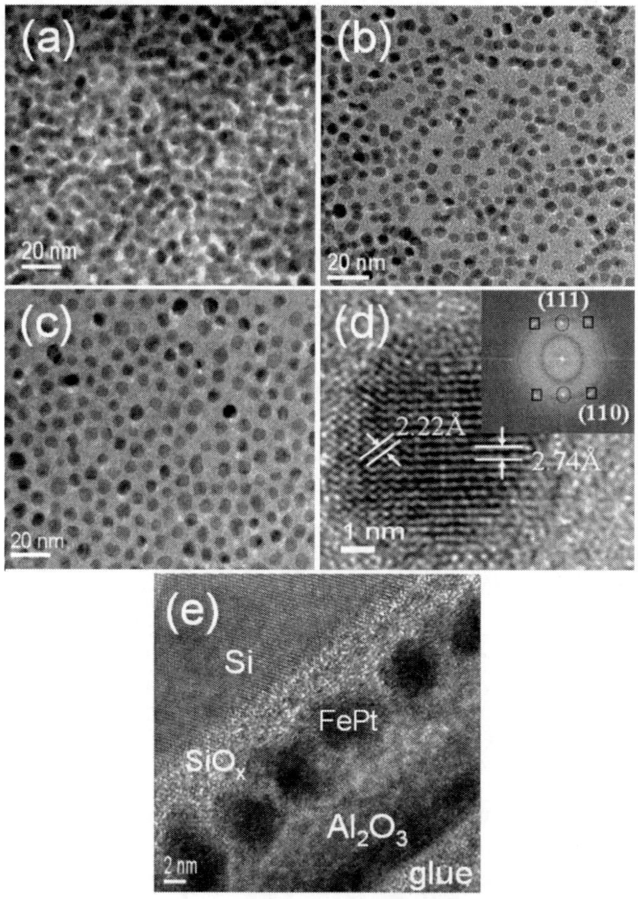

Figure 40. TEM images of FePt NCs on Si substrates after annealing at 700 °C for 1 h: (a) without Al_2O_3 protection layer; (b) and (c) with 5 nm-thick and 10 nm-thick Al_2O_3 capping layer, respectively; (d) the plan-view HRTEM image of the 5 nm fct-FePt NCs/10 nm-thick Al_2O_3-matrix composite thin film, Inset (d) is the corresponding fast Fourier transform (FFT) result; (e) the cross-sectional HRTEM image of the FePt-NC/10 nm-thick Al_2O_3-matrix composite film after annealing.

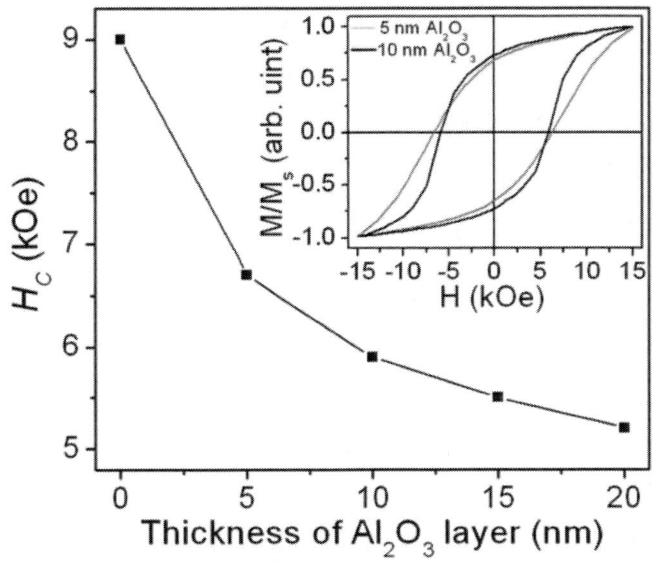

Figure 41. Plot of coercivity H_C as a function of the Al_2O_3 layer thickness. The inset is room-temperature hysteresis loops of fct-FePt NCs/Al_2O_3-matrix samples with Al_2O_3 layer thickness of 5, and 10 nm, respectively. (Reprinted with permission from [31]; copyright 2011, Royal Society of Chemistry)

An approximate relation between degree of the chemical ordering parameter S and c/a (where c and a are the lattice constants for the fct-FePt, evaluated from the (001) and (110) Bragg peaks of the XRD patterns) can be written as follows [69, 72]:

$$S^2 = \frac{1-(c/a)_{measured}}{1-(c/a)_{complete}}$$

where $(c/a)_{complete} = 0.9657$ is the axial ratio for the fully ordered-phase FePt, $(c/a)_{measured}$ is the axial ratio for the partially ordered phase and can be evaluated from the superlattice and fundamental peaks in the XRD patterns. The chemical ordering parameter $S = 1$ in this equation corresponds to the fully ordered phase. The long-range S values are 0.98, 0.96, 0.92 and 0.86 for the sample without Al_2O_3 protection layer, and the samples with 5 nm-thick, 10 nm-thick and 20 nm-thick Al_2O_3 capping layers, respectively. The slight decrease of the chemical ordering parameter S is attributed to the smaller FePt particle size embedded in Al_2O_3-matrix and the residual strain at the FePt/Al_2O_3 interfaces. Evidently, the fcc to fct phase transition is more difficult for FePt NCs with smaller size (close to critical size).

TEM image in Figure 40 (a) shows obvious grain growth and aggregation of FePt NCs in samples without any protection layer. After the same 700 °C annealing process, the sample with 5 nm Al_2O_3 capping layer displays evidently improved dispersion of FePt NCs (Figure 40 (b)), but small amount of aggregation can still be observed. With further increasing the Al_2O_3 layer to 10 nm, the well-separated FePt NCs embedded in the amorphous Al_2O_3-matrix layer without significant morphology change are confirmed in Figure 40 (c). Figure 40 (d) shows a representative HRTEM image of one FePt NC in the 10 nm-thick Al_2O_3 capping layer. The (110) and (111) lattice planes can be well recognized with lattice spacing of 2.74 Å and 2.22 Å, larger than that of bulk FePt, in good accordance with the XRD results. The corresponding fast Fourier transform (FFT) indicates the $L1_0$ structure of the FePt NCs. The cross-sectional HRTEM image of the 5 nm fct-FePt NCs/10 nm-thick Al_2O_3 composite film after annealing has been recorded in Figure 40 (e). It is clearly seen that a monolayer of FePt NCs is uniformly embedded in amorphous SiO_x and Al_2O_3 matrix. It is worth noting that the top amorphous Al_2O_3 capping layer is very flat and homogeneous with 10 nm-thick. Above all, the chemically stable amorphous layers can serve as an effective barrier for separating from FePt NCs.

The coercivities of the composite films as a function of the thickness of ALD Al_2O_3 layer are plotted in Figure 41. A high coercivity of 8.9 kOe is obtained for fct-FePt NC-assembled film without the Al_2O_3 capping layer. With the increase of the thickness of Al_2O_3 layer from 5 nm to 20 nm, the coercivity of the corresponding samples decreases gradually from 6.6 kOe to 5.1 kOe. Inset of Figure 41 shows the hysteresis loops of the fct-FePt NCs covered with 5 and 10 nm-thick Al_2O_3-matrix layers. It should be noted that the grain size is a key factor affecting the magnetic properties and $L1_0$ ordering process [73]. The coercivity increases rapidly with the average grain or particle size [74, 75], up to a maximum single domain size, after enduring the same annealing process. FePt NCs assembly without Al_2O_3 capping layer show significant grain growth and aggregation with an average grain size of 14.9 nm from XRD data, leading to the enhancement of coercivity. Meanwhile, the decrease of coercivity for FePt NCs with Al_2O_3 capping layer is related to the reduced chemical ordering S and the presence of residual interfacial strain [76] between FePt NCs and Al_2O_3 capping layer. When the thickness of amorphous Al_2O_3 capping layer is 10 nm, the nanocomposite films exhibit optimal dispersion and magnetic property after high-temperature annealing. The fct-FePt NCs/10 nm-thick Al_2O_3 sample possess a coercivity of 5.9 kOe and magnetocrystalline anisotropy constant K_u of 3.86 MJ/m^3. The thermal factor η for above sample is 68.5, meeting the industry requirement ($\eta \geq 50$).

CONCLUSION

In this chapter, several novel nonvolatile memory devices, such as CTM, RRAM, FeRAM, and patterned magnetic storage, have been addressed, among which ALD functional layers play an important and unique role. The impact of ALD fabrication processing, composition, memory architecture and microstructure, and electrode of nonvolatile memory devices on the data storage characteristics has been discussed deeply. For CTM devices, flexible ALD is utilized to deposit tunneling/blocking layer and CTL such as NC or high-*k* composite materials. A series of metal NC/high-*k* dielectric NC memory and new SONOS-type CTM cells are presented and compared based on their charge trapping capability. The optimal CTM structure and processing has been obtained with larger memory window, faster P/E response, and better retention property. ALD-derived trilayer structure of $Al_2O_3/HfO_2/Al_2O_3$ and Al_2O_3 based $CoPt_x$ NC RRAM cells on TiN bottom electrodes have been first prepared, exhibiting reliable and reproducible resistive switching behaviors with narrow distribution of resistive switching parameters. Zr, Al, Si or Y doped Hf-based oxide ferroelectric thin films have been explored by ALD for FeRAM applications. The origin of ferroelectricity of Hf-based oxide ultrathin films has been proposed. Finally, the combination of ALD-capping layer and self-assembled FePt NCs provides a new potential approach to fabricate patterned magnetic recording media with ultrahigh areal density (50Tbit/inch^2). All these indicate that ALD technique is the most suitable and necessary method for depositing the critical layers in the nonvolatile memory devices with tremendous prospects due to its conformality, uniformity, precise and easy control of film thickness, and low processing temperature. This work described in this chapter helps to promoting ALD practical applications in next-generation nonvolatile memories in the future.

ACKNOWLEDGMENTS

This work is supported by the Natural Science Foundation of China (51202107), a grant from the State Key Program for Basic Research of China (2015CB921203 and 2011CB922104), Doctoral Fund of Ministry of Education of China (20120091110049), and the PAPD in Jiangsu Province. The author also thanks her Ph.D. students of Xiao-Jie Liu, Ji-Zhou Kong and Lai-Guo Wang, M.S. student student of Zheng-Yi Cao for their work and her colleagues of Profs. Jiang Yin and Zhi-Guo Liu for their helpful cooperation.

REFERENCES

[1] Packan, A. *Science*, 285, 2079 (1999).

[2] Kingon, AI; Maria, JP; Streiffer, SK. *Nature*, 46 (2000) 1032.

[3] *The International Technology Roadmap for Semiconductors*, Semiconductor Industry Association, USA (2001, 2007, 2009, 2013), http://www.itrs.net/ reports.html

[4] Schumacher, M; Baumann, PK; Seidel, T. *Chem. Vap. Dep.*, 12, 99 (2006).

[5] Meena, JS; Sze, SM; Chand, U; Tseng, TY. *Nano. Res. Lett.*, 9, 526 (2014).

[6] Chung, A; Deen, J; Lee, JS; Meyyappan, V. *Nanotechnology*, 21, 412001 (2010).

[7] Chang, TC; Jian, FY; Chen, SC; Tsai, YT. *Mater. Today*, 14, 608 (2011).

[8] Pan, F; Gao, S; Chen, C; Song, C; Zeng, F. *Mater. Sci. Eng.*, R 83, 1 (2014).

[9] Scott, JF; de Araujo, CAP. *Science*, 246, 1400 (1989).

[10] Böscke, TS; Müller, J; Bräuhaus, D; Schröder, U; Böttger, U. *Appl. Phys. Lett.*, 99, 102903 (2011).

[11] Xu, YF; Yan, ML; Sellmyer, DJ. *J. Nanosci. Nanotech.*, 7, 206 (2007).

[12] Sherman, A. *Atomic layer deposition for nanotechnology*, Ivoryton Press, Ivoryton (2008).

[13] Pinna, N; Knez, M. *Atomic Layer Deposition of Nanostructured Materials*, Wiley-VCH Verlag & Co. KGaA, Weinheim (2012).

[14] George, SM. *Chem. Rev.*, 110, 111 (2010).

[15] Ritala, M; Niinistő, J. *Chap. 4 Atomic Layer Deposition* in *Chemical Vapour Deposition Precursors, Processes and Applications*, Royal Society of Chemistry, Cambridge (2009).

[16] Kahng, D; Sze, SM. *Bell Syst. Tech. J.*, 46, 1288 (1967).

[17] White, MH; Adams, DA; Bu, JK. *IEEE Circuits and Devices*, 16, 22 (2000).

[18] Tiwari, S; Rana, F; Hanafi, H; Hartstein, A; Crabbé, EF; Chan, K. *Appl. Phys. Lett.*, 68, 1377 (1996).

[19] Lee, C; Meteer, J; Narayanan, V; Kan, EC. *J. Electron. Mater.*, 34, 1 (2005).

[20] Liu, XJ; Zhu, L; Gao, MY; Li, XF; Cao, ZY; Zhai, HF; Li, AD; Wu, D. *Appl. Surf. Sci.*, 289, 332 (2014).

[21] Liu, XJ; Zhu, L; Li, XF; Cao, ZY; Li, AD; Wu, D. *J. Vac. Sci. Technol. B*, 2, 042201 (2014).

[22] Lan, XX; Ou, X; Xu, B; Gong, CJ; Li, R; Yin, QN; Xia, YD; Yin, J; Liu, ZG; Li, AD; Yan, F. *Chin. Phys. Lett.*, 30, 128102 (2013).

[23] Liu, XJ; Gao, MY; Li, AD; Cao, YQ; Li, XF; Guo, BL; Cao, ZY; Wu, D. *J. Alloys Compd.*, 588, 103 (2014).

[24] Lee, HBR; Mullings, MN; Jiang, X; Clemens, BM; Bent, SF. *Chem. Mater.*, 24, 4051 (2012).

[25] Yang, HY; Niimi, H; Lucovsky, G. *J. Appl. Phys.*, 83, 2327 (1998).

[26] Lu, TZ; Alexe, M; Scholz, R; Talelaev, V; Zacharias, M. *Appl. Phys. Lett.*, 87, 202110 (2005).

[27] Liu, XJ. *Investigation on fabrication and memory effect of metal nanocrystal charge trapping memory cells*, PhD dissertation thesis, Nanjing Univ., (2015). pp. 44–99.

[28] Kim, JK; Cheong, HJ; Kim, Y; Yi, JY; Bark, HJ; Bang, SH; Cho, JH. *Appl. Phys. Lett.*, 82, 2527 (2003).

[29] Lwin, ZZ; Pey, KL; Liu, C; Liu, Q; Zhang, Q; Chen, YN; Singh, PK; Mahapatra, S. *Appl. Phys. Lett.*, 99, 222102 (2011).

[30] Heitmann, J; muller, F; Yi, L; Zacharias, M; Kovalev, D; Eichhom, F. *Phys. Rev. B*, 69, 195309 (2004).

[31] Kong, JZ; Gong, YP; Li, XF; Li, AD; Zhang, JL; Yan, QY; Wu, D. *J. Mater. Chem.*, 21, 5046 (2011).

[32] Kong, JZ; Gao, MY; Xia, YD; Li, AD; Zhang, JL; Gong, YP; Yan, QY; Wu, D. *J. Alloys Compd.*, 542, 128 (2012).

[33] Leu, CC; Chen, ST; Liu, FK; Wu, CX. *J. Mater. Chem.*, 22, 2089 (2012).

[34] Maikap, S ; Lee, HY; Wang, TY; Tzeng, PJ; Wang, CC; Lee, LS; Liu, KC; Yang, JR; Tsai, MJ. *Semicond. Sci. Technol.*, 22, 884 (2007).

[35] Lai, CH; Chin, A; Kao, HL; Chen, KM; Hong, M; Kwo, J; Chi, CC. *Dig. Tech. Pap. - Symp. VLSI Technol.*, 44 (2006).

[36] Fu, YY; Li, AD; Liu, XJ; Li, XF; Wu, D; Tang, ZJ. *Characteristics of atomic layer deposition-derived all-high-κ-based structures for flash memory application,* 11th Non-Volatile Memory Technology Symposium (NVMTS 2011), Shanghai, China, Nov. 7-9, 2011.

[37] Cao, ZY; Li, AD; Li, X; Cao, YQ; Wu, D. *Thin Solid Films*, 563, 6 (2014).

[38] Zhou, Y; Yin, J; Xu, HN; Xia, YD; Liu, ZG; Li, AD; Gong, YP; Pu, L; Yan, F; Shi, Y. *Appl. Phys. Lett.*, 97, 143504 (2010).

[39] Tang, ZJ; Zhu, XH; Xu, HN; Xia, YD; Yin, J; Li, AD; Yan, F; Liu, ZG. *Appl. Phys. A*, 108, 217 (2012).

[40] Chau, R; Doyle, B; Datta, S; Kavalieros, J; Zhang, K. *Nat. Mater.*, 6, 810 (2007).

[41] Tang, ZJ; Xia, YD; Xu, HN; Yin, J; Liu, ZG; Li, AD; Liu, XJ; Yan, F; Ji, XL. *Electrochem. Solid-State Lett.*, 14, G13 (2011).

[42] Lan, XX; Gong, CJ; Ou, X; Cao, YQ; Sun, C; Chen, Y; Xu, B; Xia, YD; Li, AD; Yin, J; Liu, ZG. *Microelectron. Eng.*, 133, 88 (2015).
[43] Tang, ZJ; Zhu, XH; Xu, HN; Xia, YD; Yin, J; Li, AD; Yan, F; Liu, ZG. *Mater. Lett.*, 92, 21 (2013).
[44] Lan, XX; Ou, X; Cao, YQ; Tang, SY; Gong, CJ; Xu, B; Xia, YD; Yin, J; Li, AD; Yan, F; Liu, ZG. *J. Appl. Phys.*, 114, 044104 (2013).
[45] Jiang, K; Ou, X; Lan, XX; Cao, ZY; Liu, XJ; Lu, W; Gong, CJ; Xu, B; Li, AD; Xia, YD; Yin, J; Liu, ZG. *Appl. Phys. Lett.*, 104, 263506 (2014).
[46] Gong, CJ; Yin, QN; Ou, X; Lan, XX; Liu, JQ; Sun, C; Wang, LG; Lu, W; Yin, J; Xu, B; Xia, YD; Liu, ZG; Li, AD. *Appl. Phys. Lett.*, 105, 123504 (2014).
[47] Yang, JJ; Strukov, DB; Stewart, DR. *Nat. Nanotechnol.*, 8, 13 (2013).
[48] Lee, J; Bourim, EM; Lee, W; Park, J; Jo, M; Jung, S. *Appl. Phys. Lett.*, 97, 172105 (2010).
[49] Waser, R; Dittmann, R; Staikov, G; Szot, K. *Adv. Mater.*, 21, 2632 (2009).
[50] Wang, LG; Qian, X; Cao, YQ; Cao, ZY; Fang, GY; Li, AD; Wu, D. *Nano. Res. Lett.*, 10, 135 (2015).
[51] Cao, CY. *Study on fabrication and memory characteristics of new type nonvolatile memory cells by atomic layer deposition*, Master's dissertation thesis, Nanjing Univ. (2015). pp. 53–68.
[52] Wang, LG; Cao, ZY; Qian, X; Zhai, HF; Li, AD; Wu, D. *Enhanced resistive switching characteristics and mechanism in TiN/CoPt$_x$/Al$_2$O$_3$/Pt nanocrystal resistive random access memory devices*, 2015 CMRS Meeting H session, July 10-14, Guiyang, China, 2015.
[53] Araujo, CA; Cuchlaro, JD; McMillan, LD; Scott, MC; Scott, JF. *Nature*, 374, 627 (1995).
[54] Kim, K; Lee, S. *J. Appl. Phys.*, 100, 51604 (2006).
[55] Bohr, MT; Chan, RS; Ghani, T; Mistry, K. *IEEE Spectrum*, 44, 29 (2007).
[56] Müller, J; Böscke, TS; Bräuhaus, D; Schröder, U; Böttger, U; Sundqvist, J; Köucher, P; Mikolajick, T; Frey, L. *Appl. Phys. Lett.*, 99, 112901 (2011).
[57] Müller, J; Schröder, U; Böscke, TS; Müller, I;Böottger, U; Wilde, L; Sundqvist, J; Lemberger, M; Köucher, P; Mikolajick, T; Frey, L. *J. Appl. Phys.*, 110, 114113 (2011).
[58] Mueller, S; Müeller, J; Singh, A; Riedel, S; Sundqvist, J; Schöder, UU; Mikolajick, T. *Adv. Funct. Mater.*, 22, 2412 (2012).
[59] Müller, J; Böscke, TS; Schöder, U; Mueller, S; Bräuhaus, D; Böttger, U; Frey, L; Mikolajick, T. *Nano Lett.*, 12, 4318 (2012).
[60] Park, MH; Kim, HJ; Kim, YJ; Lee, W; Moon, T; Hwang, CS. *Appl. Phys. Lett.*, 102, 242905 (2013).
[61] Clima, S; Wouters, DJ; Adelmann, C; Schenk, T; Schroeder, U; Jurczak, M; Pourtois, G. *Appl. Phys. Lett.*, 104, 092906 (2014).
[62] Park, MH; Kim, HJ; Kim, YJ; Lee, W; Kim, HK; Hwang, CS. *Appl. Phys. Lett.*, 102, 112914 (2013).
[63] Mueller, S; Müller, J; Schroeder, U; Mikolajick, T. *IEEE Trans. Device Mater Reliability*, 13, 93 (2013).
[64] Mueller, S; Summerfelt, S; Müller, J; Schröder, U; Mikolajick, T. *IEEE Electron Device Lett.*, 33, 1300 (2012).
[65] Lee, J; Bourim, EM; Lee, W; Park, J; Jo, M; Jung, S. *Appl Phys Lett*, 97, 172105 (2010).

[66] Sun, SH. *Adv. Mater.*, 18, 393 (2006).
[67] Zeng, H; Li, J; Liu, JP; Wang, ZL; Sun, SH. *Nature*, 420, 395 (2002).
[68] Sun, SH; Murray, CB; Weller, D; Folks, L; Moser, A. *Science*, 287, 1989 (2000).
[69] Yan, Q; Kim, T; Purkayastha, A; Ganesan, PG; Shima, M; Ramanath, G. *Adv. Mater.*, 17, 2233 (2005).
[70] Yan, Q; Kim, T; Purkayastha, A; Xu, Y; Shima, M; Gambino, RJ; Ramanath, G. *J. Appl. Phys.*, 99, 08N709 (2006).
[71] Kong, JZ; Gao, MY; Zhai, HF; Yan, QY; Li, AD; Li, H; Wu, D. *J. Mag. Mag. Mater.*, 343, 1 (2013).
[72] Nandwana, V; Elkins, KE; Poudyal, N; Chaubey, GS; Yano, K; Liu, JP. *J. Phys. Chem. C*, 111, 4185 (2007).
[73] Ding, Y; Majetich, SA. *Appl. Phys. Lett.*, 87, 022508 (2005).
[74] Jeong, U; Teng, XW; Wang, Y; Yang, H; Xia, YN. *Adv. Mater.*, 19, 33 (2007).
[75] Ding, Y; Farrell, D; Yamamuro, S; Majetich, SA. *J. Appl. Phys.*, 93, 7411 (2003).
[76] Takahashi, YK; Hono, K. *Appl. Phys. Lett.*, 84, 383 (2004).

In: Atomic Layer Deposition (ALD)
Editor: Jeannie Valdez
ISBN: 978-1-63483-869-6

Chapter 4

THEORETICAL CALCULATION ON MECHANISM OF THERMAL AND PLASMA-ENHANCED ATOMIC LAYER DEPOSITION OF SiO_2

Guo-Yong Fang[1,2,*], Li-Na Xu[2] and Ai-Dong Li[1,†]
[1]National Laboratory of Solid State Microstructures, College of Engineering and Applied Sciences, Collaborative Innovation Center of Advanced Microstructures, Nanjing University, Nanjing, China
[2]Zhejiang Provincial Key Laboratory of Carbon Materials, College of Chemistry and Materials Engineering, Wenzhou University, Wenzhou, China

ABSTRACT

Due to the excellent physical and chemical properties, thin films of silicon dioxide (SiO_2) have played a very important role in microelectronics, photonics, catalysis, and energy. At the same time, atomic layer deposition (ALD) is a powerful deposition technique and can control the composition and thickness of thin films at the atomic scale. This chapter summarizes our theoretical works on the reaction mechanisms of thermal and plasma-enhanced ALD (PE-ALD) of SiO_2 by using density functional theory calculations. Thermal ALD of SiO_2 includes H_2O-based ALD without a catalyst, room-temperature ALD (RT-ALD) with Lewis base catalyst, and rapid ALD (RALD) with Lewis acid catalyst. In H_2O-based ALD of SiO_2, the stepwise mechanism and H_2O-assisted hydrolysis were found. In RT-ALD of SiO_2, the surface pseudorotation (SPR) phenomenon and Lewis-base catalyzed multi-step pathway, including adsorption of precursor, axial addition, surface pseudorotation, axial elimination and desorption of byproduct, were explored. In RALD of SiO_2, the synergistic catalysis of Lewis/Brønsted acid sites and interfacial interactions, which accelerates the growth of silica nanolaminate, was also analyzed. In PE-ALD of SiO_2, the concepts of self-catalysis by aminosilanes and strong surface oxidation by O_2 plasma were proposed. All these

* E-mail address: fanggy@wzu.edu.cn.
† E-mail address: adli@nju.edu.cn.

indicate that surface reactions of SiO_2 ALD are complex and may be exploited to improve ALD preparation of other oxides and their nanolaminates, such as Al_2O_3, HfO_2, TiO_2, Al_2O_3/SiO_2, and HfO_2/SiO_2.

INTRODUCTION

Due to the excellent physical and chemical properties, thin films of silicon dioxide (SiO_2) have played a very important role in microelectronics, photonics, catalysis, and energy [1-3]. Usually, SiO_2 thin films can be prepared by various methods, such as thermal oxidation of silicon, sol-gel method (Sol-Gel), physical vapor deposition (PVD), chemical vapor deposition (CVD), and atomic layer deposition (ALD). Among these techniques, ALD allows accurate control of the thickness and composition of thin films at the atomic scale and is an increasingly popular alternative to fabricate ultrathin uniform and conformal films [4].

Essentially, ALD is a special CVD technique. CVD is a continuous process where all reactants are simultaneously supplied in the reaction chamber to grow the film, but ALD divides the CVD reaction into two half-reactions and separately introduces the precursors into the reaction chamber. For example, SiO_2 ALD, using silicon tetrachloride ($SiCl_4$) and water (H_2O), divides CVD reaction into two individual half-reactions, **A** and **B**,

CVD: $SiCl_4 + 2H_2O \rightarrow SiO_2 + 4HCl$,
ALD: (**A**) $Si\text{-}OH^* + SiCl_4 \rightarrow Si\text{-}O\text{-}SiCl^* + HCl$,
(**B**) $Si\text{-}O\text{-}SiCl^* + H_2O \rightarrow Si\text{-}O\text{-}Si\text{-}OH^* + HCl$,

where an asterisk designates the surface species.

In general, SiO_2 ALD requires high reaction temperatures and large precursor fluxes, and that is very slow [5-7]. Experimentally, the room-temperature deposition of SiO_2 can be achieved by introducing Lewis base catalyst, such as pyridine (NC_5H_5) or ammonia (NH_3), named as room-temperature ALD (RT-ALD) [8, 9]. The aminosilanes with amino ligands have similar catalytic role, but subsequent half-reaction needs ozone (O_3) or O_2 plasma as oxidant [10]. If Lewis acid catalyst, such as trimethylaluminum (TMA, $Al(CH_3)_3$), is used, the growth rate of silica can be increased 100 times compared with conventional ALD [11]. This deposition is named as rapid ALD (RALD). By using density functional theory calculations, we have studied above mentioned depositions of SiO_2.

This chapter summarizes our theoretical works on the reaction mechanisms of thermal and plasma-enhanced ALD (PE-ALD) of SiO_2 [12-15]. We hope all these results can deepen the understanding on the SiO_2 ALD chemistry and provide theoretical guides for the design of Si precursor and ALD preparation of other oxides, such as Al_2O_3, HfO_2, TiO_2, Al_2O_3/SiO_2, and HfO_2/SiO_2.

MECHANISM OF THERMAL ALD OF SiO_2

(i) H_2O-Based ALD without a Catalyst

Various experimental investigations have shown that $SiCl_4$ and H_2O half-reactions result in chlorinated and hydroxylated surfaces, respectively [5-7]. In order to get more insight into the reaction mechanism of H_2O-based ALD of SiO_2 without a catalyst, theoretical calculation has been performed that illustrate the reaction pathways [16].

It was proposed that on the Si(001) surface, the half-reaction of $SiCl_4$ with surface hydroxyl group (-OH) proceeds through a concerted pathway via a four-membered ring (4MR) transition state (TS), forming Si-O and H-Cl bonds while simultaneously breaking Si-Cl and O-H bonds [16]. Unlike other Lewis acids, such as $AlCl_3$, $TiCl_4$, $ZrCl_4$, and, $HfCl_4$, through chemisorption to form metal-oxygen bond directly [11, 17], however, $SiCl_4$ seems to have no strong nucleophilicity to attack the O atom and form Si-O bond.

Recently, the reaction mechanism of the full cycle of SiO_2 ALD on the actual SiO_2 surface was investigated by using detailed density functional theory (DFT) calculations [12], involving the $SiCl_4$ half-reaction and H_2O half-reaction (Figure 1).

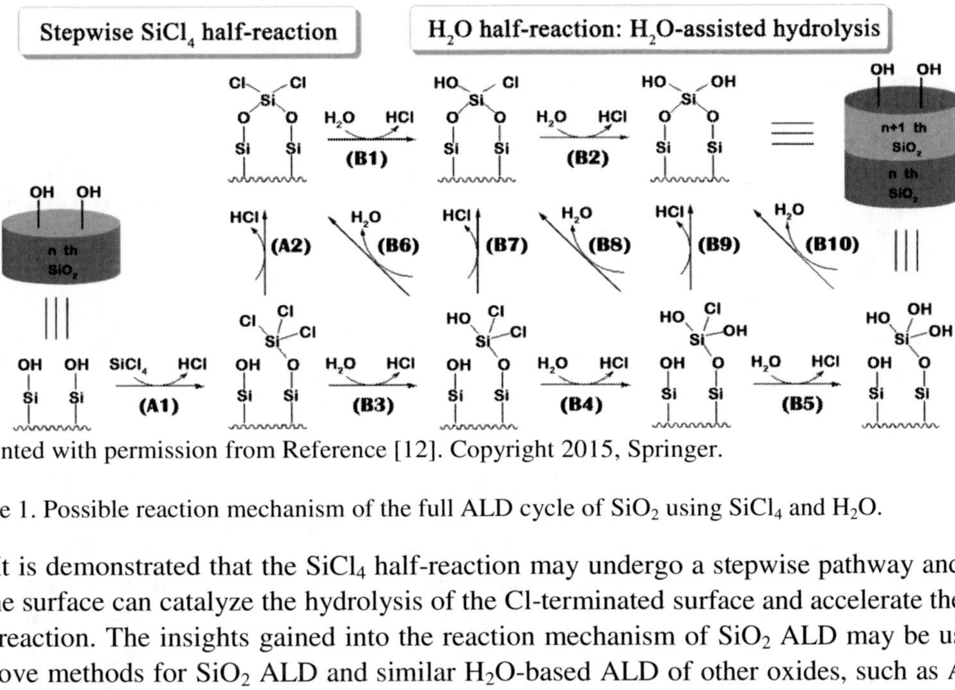

Reprinted with permission from Reference [12]. Copyright 2015, Springer.

Figure 1. Possible reaction mechanism of the full ALD cycle of SiO_2 using $SiCl_4$ and H_2O.

It is demonstrated that the $SiCl_4$ half-reaction may undergo a stepwise pathway and H_2O on the surface can catalyze the hydrolysis of the Cl-terminated surface and accelerate the H_2O half-reaction. The insights gained into the reaction mechanism of SiO_2 ALD may be used to improve methods for SiO_2 ALD and similar H_2O-based ALD of other oxides, such as Al_2O_3, TiO_2, ZrO_2, and HfO_2.

(A) $SiCl_4$ Half-Reaction: Stepwise Mechanism

Figure 2 shows the reaction pathway for the $SiCl_4$ half-reaction between $SiCl_4$ precursor and the surface hydroxyl (-OH) on the surface. Firstly, reaction **A1** between $SiCl_4$ and -OH undergoes a stepwise pathway and goes through a rotation transition state, $\mathbf{TS1^{A1}}$, and simultaneously forms a pentacoordinated intermediate, $\mathbf{Im2^{A1}}$. Subsequently, the unstable intermediate undergoes the second transition state, $\mathbf{TS2^{A1}}$, forming the product -$OSiCl_3^*$, $\mathbf{P^{A1}}$ and accompanied by the release of HCl.

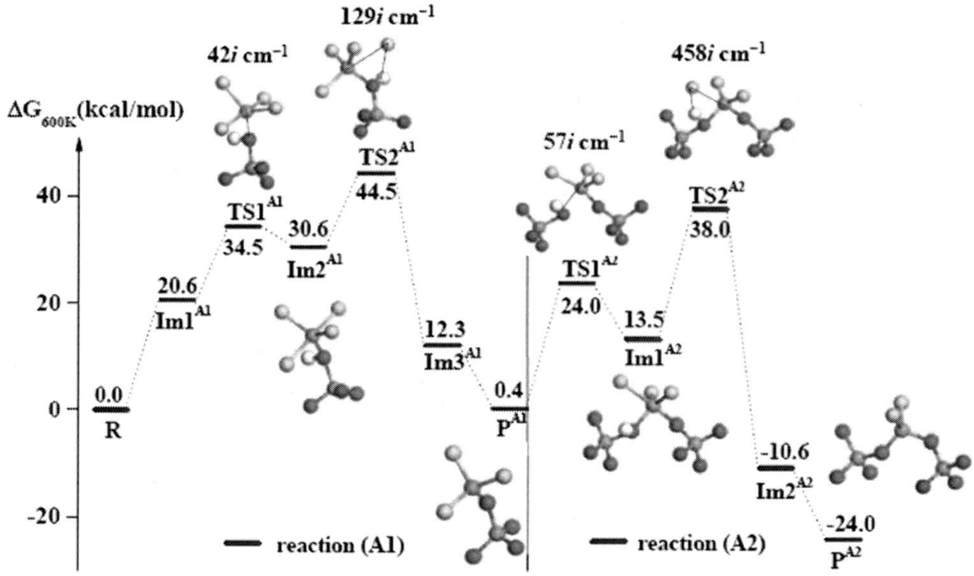

Figure 2. Gibbs free energy profile of the $SiCl_4$ half-reaction of SiO_2 ALD.

Due to massive -OH on the SiO_2 surface after H_2O half-reaction, -$OSiCl_3^*$ on the surface can further react with another adjacent -OH to form the bridged product -$O_2SiCl_2^*$, $\mathbf{P^{A2}}$. Similar to the stepwise reaction **A1**, reaction **A2** between -$OSiCl_3^*$ and -OH also undergoes two transition states, $\mathbf{TS1^{A2}}$ and $\mathbf{TS2^{A2}}$, and a pentacoordinated intermediate, $\mathbf{Im1^{A2}}$. The overall $SiCl_4$ half-reaction is exergonic by 24.0 kcal mol^{-1}. The highest activation free energy of the $SiCl_4$ half-reaction is 44.5 kcal mol^{-1} ($\mathbf{TS2^{A1}}$), indicating that the $SiCl_4$ half-reaction is very difficult. This difficulty can be overcome by the introduction of Lewis base catalysts, such as ammonia, pyridine, and aminosilane [8-10].

(B) H_2O Half-Reaction: H_2O-Assisted Hydrolysis

In SiO_2 CVD, $SiCl_4$ and H_2O are introduced into the reaction chamber simultaneously and form SiO_2 through hydrolysis and condensation processes. In SiO_2 ALD, $SiCl_4$ and H_2O are separately introduced into the chamber, but the hydrolysis and condensation still occur. In fact, the half-reaction between H_2O and the Cl-terminated surface exchanges Cl and -OH ligands and changes Si-Cl* species into Si-OH* species. Due to this the possible reactions of the H_2O half-reaction (**B**) may include the formation of silanol (-Si-OH) via the exchange of ligands between Cl and -OH (reactions **B1**, **B2**, **B3**, **B4**, and **B5**) and the formation of -O-Si-O- bridges by removing H_2O (reactions **B6**, **B8**, and **B10**) and HCl (reactions **B7** and **B9**), shown in Figure 1, similar to the hydrolysis (-Si-OH) and condensation (-O-Si-O-) processes of SiO_2 CVD.

In H_2O half-reaction, H_2O molecule can be adsorbed on the surface and accelerate the hydrolysis of Cl-terminated surface species, such as -O_2Si-Cl_2^*, -O_2SiOH-Cl^*, -OSi-Cl_3^*, -$OSiOH$-Cl_2^*, and -$OSi(OH)_2$-Cl^*, termed as H_2O-assisted hydrolysis, which results from the interaction of $H_2O...H_2O$ hydrogen bonding. Therefore, the $SiCl_4$ half-reaction is the rate-determining step of the full ALD cycle of SiO_2 and controls the ALD growth of SiO_2. We anticipate that H_2O-assisted hydrolysis may also exist in H_2O-based ALD of other oxides.

(ii) RT-ALD with Lewis Base Catalyst

To lower the deposition temperature to room temperature, Lewis base catalysts, such as pyridine (NC_5H_5) and ammonia (NH_3), have been used to catalyze the SiO_2 ALD [8, 9]. The initial growth of NH_3-catalyzed room temperature ALD (RT-ALD) of SiO_2 was investigated by ab initio calculation, qualitatively interpreting the role of Lewis base catalyst, which can reduce the activation energy and reaction temperature [18]. Influence of alkalinity and steric hindrance of Lewis base catalysts on SiO_2 ALD was also discussed [19].

Recently, the combination method of static transition-state search and Born-Oppenheimer molecular dynamics (BOMD) simulation has been used to investigate possible reaction pathways for Lewis base catalyzed ALD of SiO_2 [13]. As shown in Figure 3, the catalyzed reaction may undergo complicated multistep pathways, including adsorption of precursor, axial addition, surface pseudorotation, axial elimination, and desorption of byproduct steps.

Furthermore, a pseudorotation phenomenon on the surface, named as surface pseudorotation (SPR), was explored. The surface pseudorotation is similar to Berry pseudorotation (BPR), which has been proposed to explain the rapid stereomutation and fluxional behavior of the pentacoordinated main-group element and transition-metal compounds with the trigonal bipyramidal (TBP) geometry [20-22]. With one ligand pivot being linked to the surface, the catalyzed reaction possesses three possible rotation modes, corresponding to three different directions, **Path$_1$**, **Path$_2$**, and **Path$_3$** (Figure 4). The generality underlying the Berry and surface pseudorotation with the TBP topology exhibits the common fluxional behavior in the pentacoordinated compounds and stimulates the discovery of the crucial role of stereomutation in the mechanism of catalysis.

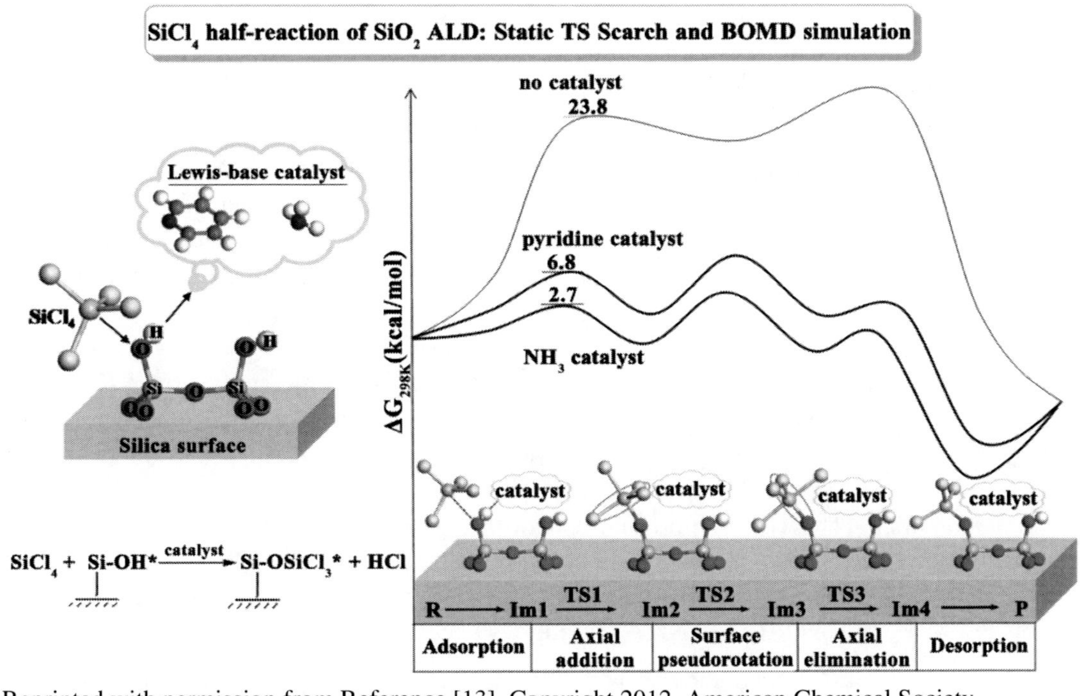

Reprinted with permission from Reference [13]. Copyright 2012, American Chemical Society.

Figure 3. Schematic illustrations of the $SiCl_4$ half-reaction of SiO_2 ALD catalyzed by Lewis-bases.

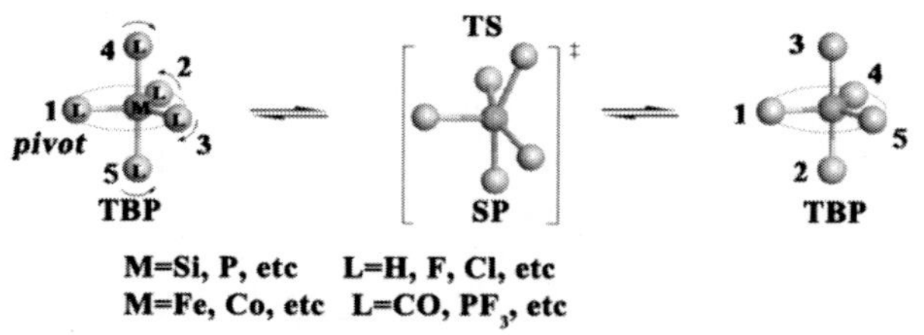

(a) Berry pseudorotation (BPR).

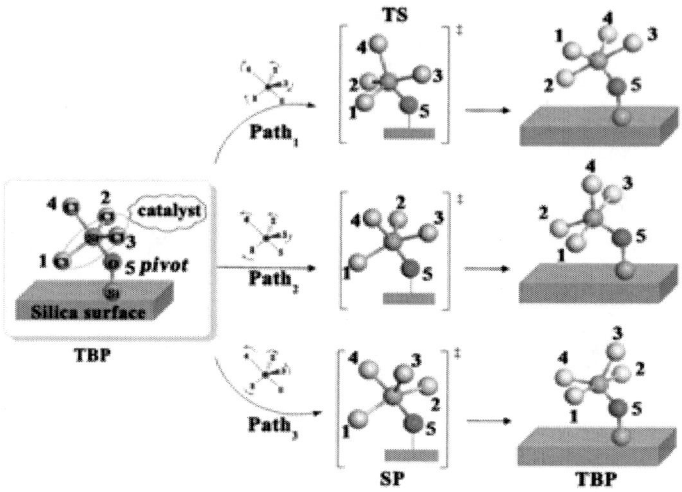

(b) Surface pseudorotation (SPR).

Figure 4. Schematic illustrations of (a) Berry pseudorotation (BPR) and (b) surface pseudorotation (SPR).

(A) Adsorption-Addition-Pseudorotation-Elimination-Desorption Pathway

As illustrated in Figures 3 and 5, whether pyridine or ammonia using as catalyst, the catalyzed reactions of SiO_2 ALD can both go through a multistep pathway, including adsorption of precursor, axial addition, surface pseudorotation, axial elimination, and desorption of byproduct.

Adsorption of precursor. In Lewis base catalyzed ALD, a steady-state of Lewis-base catalyst is maintained in the reactor. Pyridine or NH_3 prefer to interact with surface -OH* and form hydrogen bond complex. Simultaneously, $SiCl_4$ is attached to the hydroxylated surface through physisorption. The coadsorption of pyridine or NH_3 and $SiCl_4$ leads to the formation of the complex, **Im1**.

Axial addition. The coadsorption complex, **Im1**, can change into a pentacoordinated silicon intermediate, **Im2**, through a concerted transition state, **TS1**, representing the formation of Si-O bond and the elongation of O-H bond. Simultaneously, pyridine or NH_3 is protonated to form $HNC_5H_5^+$ or NH_4^+ cation. The pentacoordinated silicon intermediate ($\vdash OSiCl_4^-$) is a negatively charged compound with a TBP geometry, in which the O atom of original -OH* group acts as an axial ligand. The activation free energy of the Si-O bond formation for pyridine or NH_3 catalyst is much lower than that of the Si-O formation in the uncatalyzed reaction. Such a process of Si-O bond formation is similar to the axial addition, which was proposed in nucleophilic substitution reactions at silicon and phosphorus [23-28].

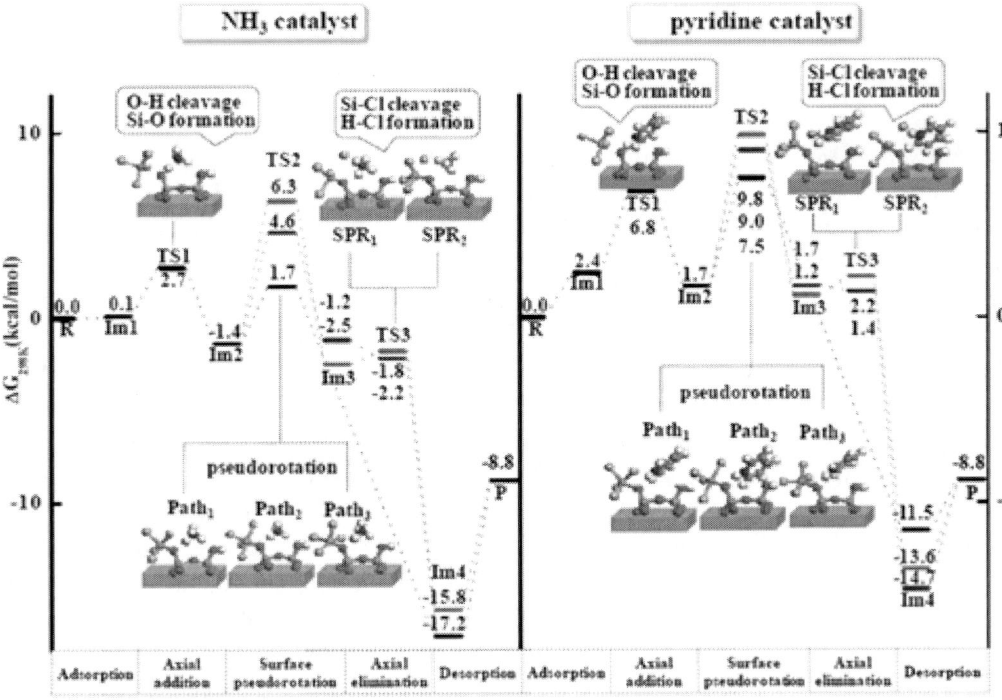

a b

Figure 5. Gibbs free energy profiles for the $SiCl_4$ half-reaction with (a) NH_3 or (b) pyridine catalyst.

Surface pseudorotation. It is possible for the pentacoordinated silicon intermediate (**Im2**) to convert into another intermediate (**Im3**) with TBP structure through pseudorotation transition states (**TS2**). Since O ligand is tied to the surface and served as the pivot, the catalyzed reaction possesses three possible rotation modes, **Path₁**, **Path₂**, and **Path₃**, respectively. From the activation barriers, the pseudorotation of the **Path₃** pathway is more favorable than other two. Due to the steric hindrance of Lewis-base cation, the pseudorotations of the **Path₁** and **Path₂** pathways, in which the axial Cl atom moves to the Lewis-base cation, have relatively higher activation energy barriers than that of **Path₃**.

Axial elimination. With the presence of the protonated Lewis-base cation ($HNC_5H_5^+$ or NH_4^+), the axial Cl atom of the pentacoordinated silicon intermediate ($\vdash OSiCl_4^-$) is further pulled away from the central silicon atom. The dissociation of axial Cl atom goes through Si-Cl cleavage transition states, **Path₁-TS3** and **Path₂-TS3**. The chloride and surface silicon compound ($\vdash OSiCl_3$) in **Im4** are subsequently formed. Although the **Path₃** is too flat to locate the transition state of the Si-Cl bond cleavage, it has the similar BOMD trajectory as **Path₁** and **Path₂** pathways. Such a process of Cl atom dissociation, Si-Cl bond cleavage and chloride formation is similar to the axial elimination in silicon and phosphorus stereochemistry [23-28]. *Desorption of byproduct*. The byproducts are ammonium chloride, pyridinium chloride and the complex between HCl and Lewis-bases. They can desorb from SiO_2 surface under a certain temperature and nitrogen pulse [8, 9].

To summarize, the whole catalyzed $SiCl_4$ half-reactions may undergo adsorption of precursor, axial addition, surface pseudorotation, axial elimination and desorption of byproduct steps. As occuring in other nucleophilic substitution reactions at silicon and phosphorus, the axial addition and elimination will be more favorable than equatorial addition and elimination [23-28].

(B) Berry Pseudorotation versus Surface Pseudorotation

The Berry pseudorotation (BPR) and the surface pseudorotation (SPR) reactions have some common features and usually occur in the pentacoordinated main-group element and transition-metal compounds with the trigonal bipyramidal (TBP) geometry. During the pseudorotation, the TBP geometry changes to another TBP geometry through via a transition state with the square pyramid (SP) geometry. Pseudorotation reactions, with or without the substrate support, are both easy to occur due to the low activation energy barriers [20-22].

The difference between Berry and surface pseudorotation reactions lies in the status of the pivot ligand. In Berry pseudorotation, the pivot atom is usually free of any spatial confinement, which is in contrast with the surface stuck pivot ligand in surface pseudorotation reaction. Thus, the surface pseudorotation is anisotropic and may have different rotation modes (Figures 4 and 6). The Berry pseudorotation results in the rapid stereomutation of TBP molecules through the pair-wise exchange of axial and equatorial ligands and the time evolution of relative energy and geometry parameters is periodic, as illustrated by the Berry pseudorotation of $SiCl_5^-$ anion and PF_5 (Figure 6a). However, the surface pseudorotation can result in the surface reactions with the other molecules, so the time evolution of relative energy is "reactive" and finally leads to the bond breaking/formation of some chemical bonds (Figure 6b). Surface pseudorotation may be facile to take place in some surface reactions containing the pentacoordinated main-group element or transition-metal compounds with the trigonal bipyramidal geometry.

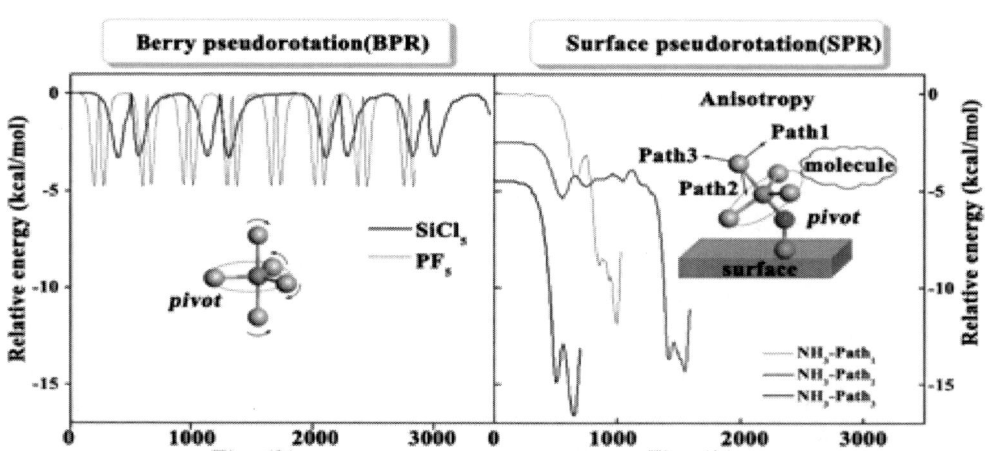

a b

Figure 6. Illustration of similarities and differences between (a) Berry pseudorotation of $SiCl_5^-$ and PF_5, and (b) surface pseudorotation of $\vdash OSiCl_4^- NH_4^+$.

(iii) RALD with Lewis Acid Catalyst

The growth rate of above mentioned ALD of SiO_2 is usually slow. Subsequently, a new deposition method using Lewis acid catalyst, such as $Al(CH_3)_3$ (TMA), namely rapid atomic layer deposition (RALD), has been developed in the microelectronic industry [11]. By the introduction of the metal and silanol precursors, silica nanolaminates were rapidly deposited at rates of more than 12 nm per ALD cycle, which is 100 times faster than the growth rate of the conventional ALD of silica [11].

In RALD, the metal and silanol precursors are introduced into the reaction chamber alternately. Similar to the conventional ALD reactions, RALD reactions also consist of two half-reactions. The first half-reaction, TMA reaction (**A**), is same as that in the conventional ALD reaction. However, the second half-reaction is more complicated than that of ALD, and it includes silanol (**B**), propagation (**C**), and cross-linking (**D**) reaction steps (Figure 7). RALD still maintains the complementary and self-limiting characteristics of the conventional ALD [14].

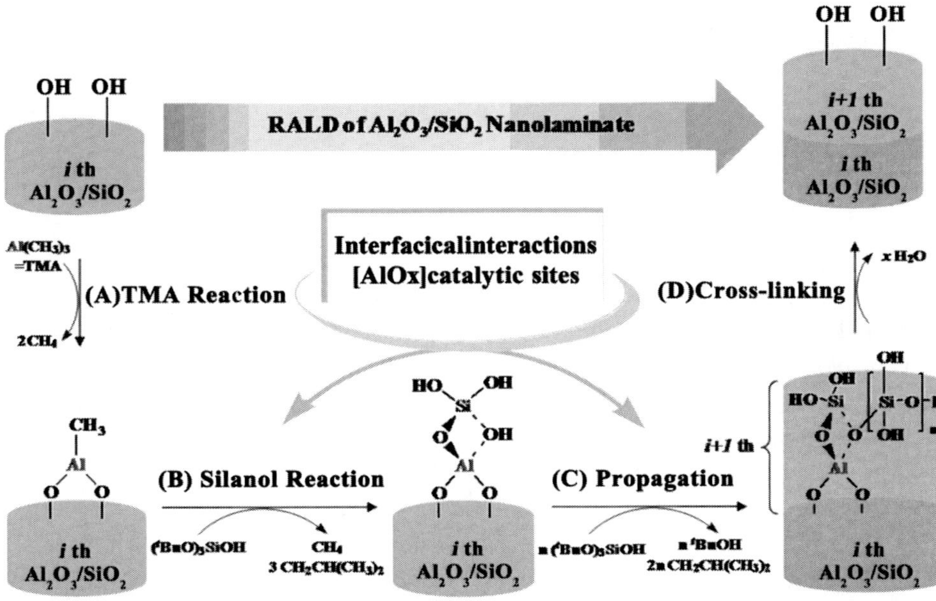

Reproduced from References [14] with permission from The Royal Society of Chemistry.

Figure 7. Schematic illustration of the rapid atomic layer deposition (RALD) of Al_2O_3/SiO_2 nanolaminate.

Recently, systematic calculations by using density functional theory (DFT) were carried out to explore the catalytic sites and possible reaction pathways for the RALD growth of Al_2O_3/SiO_2 nanolaminate [14]. The introduction of Al catalyst can form the covalently bonded $[AlO_x]$ active sites and greatly accelerate the growth of SiO_2, shown in Figure 8a. Such a catalysis reaction that takes place via the covalent formation of the activated catalyst-substrate complex is named as covalent catalysis [29]. Close to the catalytic center, some weak non-bonded interactions, such as hydrogen bonding interactions between the tert-butoxy groups and the surface, shown in Figure 8b, may also catalyze SiO_2 RALD reactions. Some references called this type of reactions as noncovalent catalysis, which are accelerated by the interactions in host-guest complexation or acid-base aggregates between catalyst and substrate [29].

The synergistic catalysis between the [AlO_x] catalytic site and interfacial interactions may be also useful in the RALD fabrication of other silica nanolaminates, such as HfO_2/SiO_2, ZrO_2/SiO_2, and TiO_2/SiO_2.

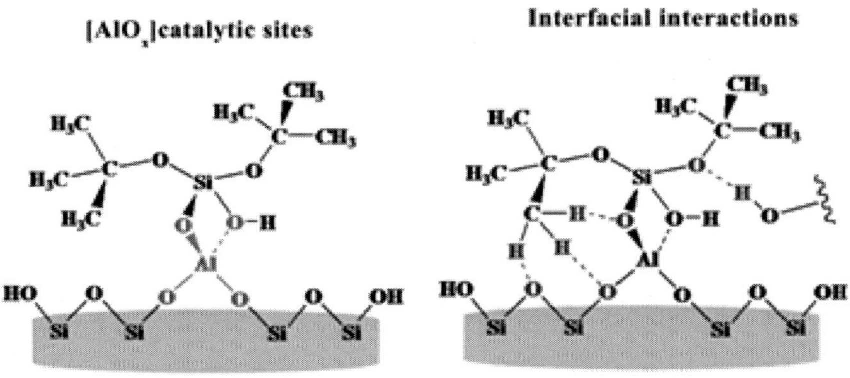

Reproduced from References [14] with permission from The Royal Society of Chemistry.

Figure 8. Schematic illustration of (a) [AlO_x] catalytic sites and (b) interfacial interactions between tert-butoxy groups and the surface.

(A) TMA Reaction

TMA adsorption and elimination of methyl. TMA processes three methyl groups, which can be deleted one by one by the release of CH_4. Firstly, TMA chemisorbs on the hydroxylated SiO_2 surface, which results in the formation of Al-O bond. Subsequently, one methyl group of the adsorbed TMA may react with one H atom of surface -OH* group to release CH_4. And then, dimethylaluminum (-$Al(CH_3)_2^*$, DMA), still bearing with two -CH_3 substituents, is attached to the surface. Similar to the release of the first CH_4, the second CH_4 can further dissociate. Finally, unimethylaluminum (–$AlCH_3^*$, UMA), with the last one -CH_3 substituent, stands on the surface, preparing for the next reaction step.

In fact, TMA, DMA, and UMA are widely used as high activity catalysts for olefins polymerizations, such as ethylene and ethylene oxide polymerizations [30-32]. Here in RALD, the catalytic activity of those alkylaluminum compounds or methyl aluminoxanes (MAO) is employed to the propagation of the siloxane into nanomaterials.

(B) Silanol Reaction

TBS adsorption and formation of [AlO_x] catalytic site. In the beginning of the second half-reaction, another precursor, TBS molecule, interacts with the MAO-planting surface. Similar to TMA adsorption, the strong interaction between TBS and the surface yields an adsorption intermediate, **Im1^B**, in which a new Al-O bond is formed. After that, the methyl group of MAO can react with H atom of hydroxyl group of TBS to release CH_4 via a transition state, **TS1^B**. Subsequently, a new aluminosilicate specie, **P1^B**, is formed with tetrahedron [AlO_4] structure and tethered to the surface (Figure 9).

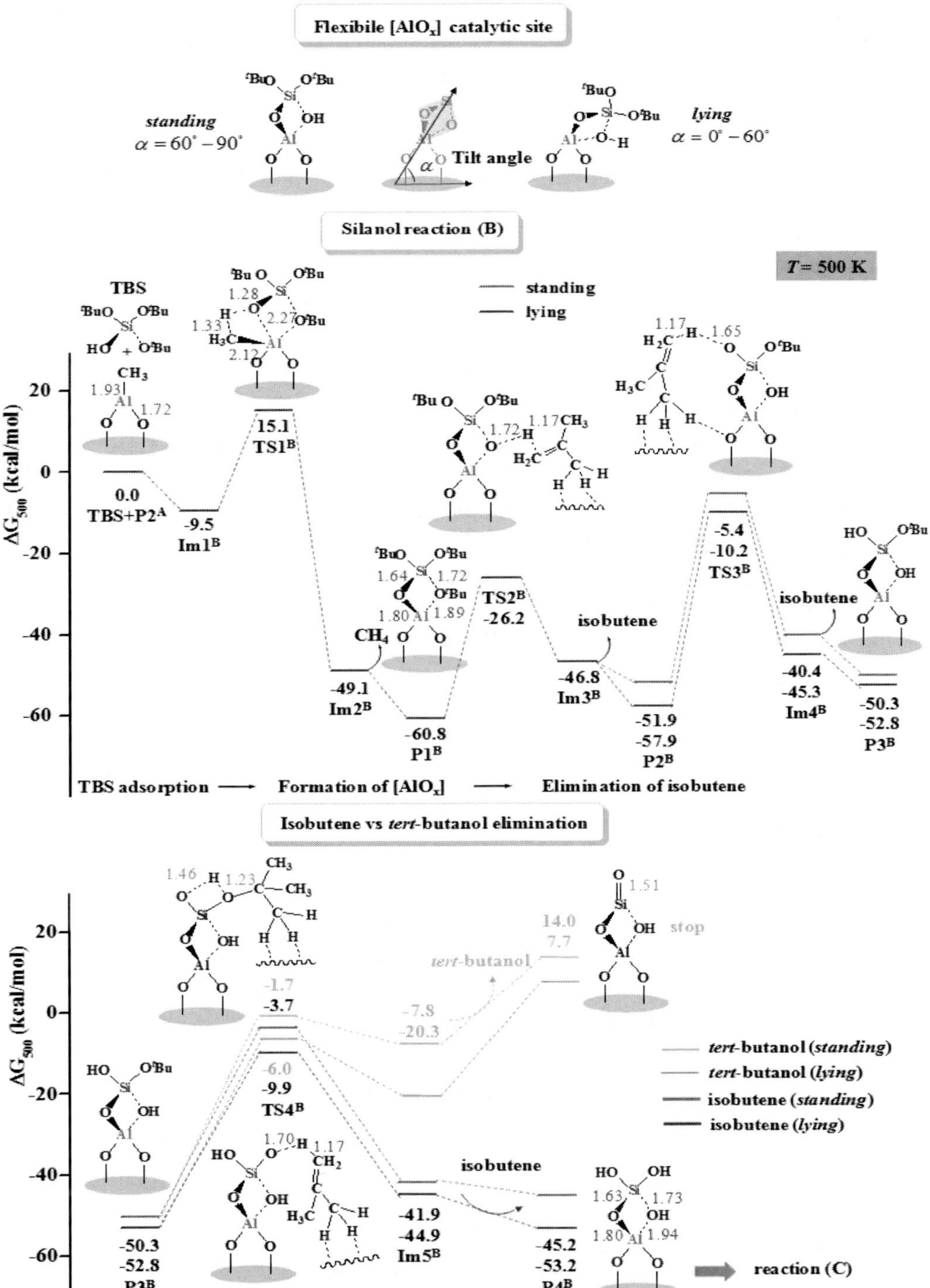

Reproduced from Reference [14] with permission from The Royal Society of Chemistry.

Figure 9. Gibbs free energy profile for the silanol reaction (B) with the Al center on the hydroxylated SiO_2 surface.

It is interesting to recall the similar four-coordinated [AlO_x] structures in zeolite and their function as the Brønsted acid site in various zeolite catalysis reactions [33-38]. Herein, the [AlO_x] site is also crucial in the catalytically elimination of isobutene.

Elimination of isobutene. At the newly formed catalytic center [AlO_4], three tert-butyl groups in the siloxane, $\mathbf{P1^B}$, will be sequentially deleted through β-hydrogen elimination of isobutene, $H_2C = CH(CH_3)_2$. Without a catalyst, the elimination of isobutene from the TBS is rather difficult with the high activation energy barrier of 61.0 kcal mol^{-1}.

However, the activation energies at the catalytic sites decrease remarkably to 32.8 kcal mol^{-1} (for the elimination of first isobutene through $\mathbf{TS2^B}$) and 45.0 kcal mol^{-1} (for the removal of the subsequent two isobutenes through $\mathbf{TS3^B}$ and $\mathbf{TS4^B}$). For overall RALD reactions, the elimination of isobutene is still the rate-determining step with the highest activation energy of about 45.0 kcal mol^{-1}.

In fact, the reaction temperature of RALD of the Al_2O_3/SiO_2 nanolaminate using TMA and TBS is about 200 ~ 300°C [11]. However, the calculated activation barrier is still too high even for the reaction at about 200 ~ 300°C. The reason may be that the present model is too idealistic relative to the real amorphous SiO_2 substrate. In addition, the coverage of reactant on surface was not considered and the intermolcular TBS···TBS interaction was also ignored. After the elimination of isobutene, the surface Si atoms are capped with three hydroxyl groups (≡Si-OH*). The new aluminosilicate species, $\mathbf{P4^B}$, is formed with similar [AlO_4] sites for catalyzing the next reaction. Different from the rigid structure of [AlO_4] in the zeolite, the tetrahedron structure [AlO_4] at Al catalytic center in RALD is rather flexible with various different orientations, *standing* and *lying* (Figure 9).

The catalytic reactivity of [AlO_4] is already well recognized in other systems. For example, in zeolite, the [AlO_4] structure has high catalytic activity in the conversion of alkenes [36-38]. But it is interesting to notice that in addition to the [AlO_4] catalytic site, the tert-butoxy groups of TBS can also react with the hydroxylated surface and serve as the catalytic sites. The interfacial interactions, including C-H···*O hydrogen bonding interactions between tert-butoxy groups and the hydroxylated surface play an important role in the elimination of isobutene. Therefore, we attempt to qualitatively partition the catalytic effects into two parts, [AlO_4] reactive sites and interfacial interactions, respectively.

The contribution from the non-bonded interactions to the lowering of activation energy can be estimated from the difference in orbital interactions between the transition state and the corresponding reactant (Figure 10). By using NBO analysis [39, 40], we can obtain the interfacial stabilization energies, $\boldsymbol{E}_2$, between the electron-donating and -accepting orbitals of the non-bonded parts. There are many electron-donating orbitals, such as O lone pair ($\boldsymbol{n}_\mathbf{O}$), O-Si bonding ($\boldsymbol{\sigma}_\mathbf{O\text{-}Si}$), O-H bonding ($\boldsymbol{\sigma}_\mathbf{O\text{-}H}$), O-Al bonding ($\boldsymbol{\sigma}_\mathbf{O\text{-}Al}$) orbitals and so on, which are able to interact effectively with the electron-accepting orbitals, the C-H antibonding $\boldsymbol{\sigma}^*_\mathbf{C\text{-}H}$ orbitals of tert-butoxy groups. The decrease of the activation energy resulting from interfacial interactions, $\boldsymbol{\Delta E}_\mathbf{a}$, is estimated from the following equation:

$$\Delta E_\mathbf{a} = E_\mathbf{2}^\mathbf{TS(i)} - E_\mathbf{2}^\mathbf{P(i\text{-}1)} \quad (1)$$

where $\boldsymbol{E}_2^{\mathrm{TS(i)}}$ and $\boldsymbol{E}_2^{\mathrm{P(i\text{-}1)}}$ represent the interfacial stabilization energies of the transition state (**TS(i)**) and the corresponding reactant (**P(i-1)**) (Figure 10). It can be found that the interfacial interactions can contribute to the reduction of activation energy, $\boldsymbol{\Delta E_a}$, of about 9 ~ 15 kcal $\mathrm{mol^{-1}}$. It is clearly shown in Figure 9 that the transition state structures favor the hydrogen bonding interactions, C-H···*O, between tert-butoxy group and the surface. Such non-bonded CH···*O interactions can stabilize the transition state and lower the activation energy barrier of the elimination of isobutene.

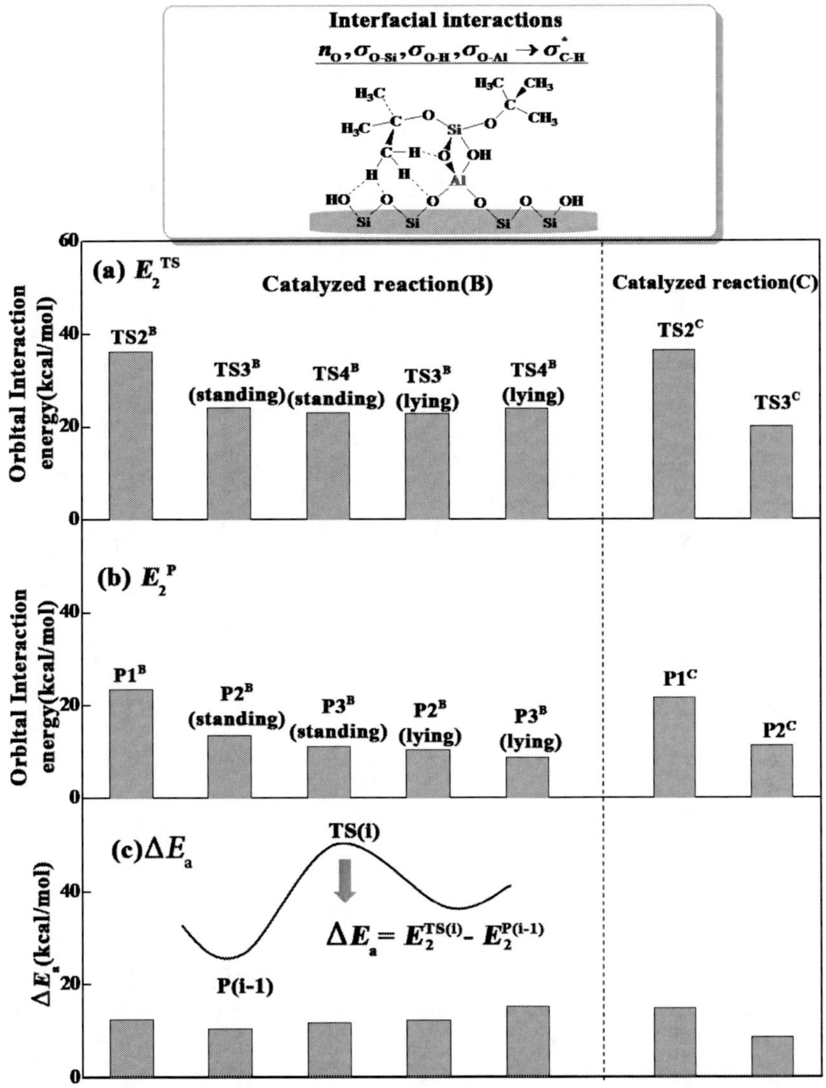

Reproduced from Reference [14] with permission from The Royal Society of Chemistry.

Figure 10. The stabilization energies (E_2, in kcal $\mathrm{mol^{-1}}$) of interfacial interactions between the C-H antibonding orbitals of tert-butoxy groups and the orbitals of surface species in (a) the transition states and (b) the corresponding reactants of the elimination of isobutene in B and C reactions, from which (c) the energy difference, ΔE_a, is calculated.

In addition to the 9 ~ 15 kcal mol^{-1} reduction of activation energy by interfacial interactions, the rest of the decrease of the activation energy can be attributed to the contribution from the [AlO_4] site. Take **TS2^B** as an example, the introduction of catalysts lowers the activation energy by 28 kcal mol^{-1}, from 61 kcal mol^{-1} (without a catalyst) to 33 kcal mol^{-1} (with the adding of TBS). The contribution of interfacial interactions to the decreasing activation energy, $\boldsymbol{\Delta E_a}$, calculated from equation (1), is about 13 kcal mol^{-1}.

As a result, the contribution of [AlO_4] to the reducing the activation energy is estimated to be 15 kcal mol^{-1}, i.e., the abstraction of 13 kcal mol^{-1} from the total contribution of 28 kcal mol^{-1}. The synergy between the covalent and noncovalent catalysis lowers the activation energy of the elimination of isobutene.

(C) The Propagation of Siloxane Polymer

TBS insertion and formation of new [AlO_4] catalytic site. In this step, another TBS molecule can attack the Al catalytic center. The insertion of TBS leads to the cleavage of one weak Al-O bond and the formation of new intermediate, **Im1^C**, in which a new Al-O bond and hence [AlO_4] catalytic site are formed (Figure 11).

Elimination of tert-butanol and formation of new Si-O bond. After the insertion of silanol into the Al-O bond, the growth of the siloxane polymer is achieved through a pentacoordinated silicon transition state (**TS1^C**), which was also found in the Lewis-base catalyzed ALD reaction [13].

Subsequently, tert-butanol molecule is released and a new Si-O bond is formed in aluminosilicate species, **P1^C**. Similar to **P4^B**, **P1^C** has the tetrahedral structure of [AlO_4]. The catalysis of the Brønsted acid site, [AlO_4], on the surface is also found in the elimination of tert-butanol through **TS1^C**. The activation energy of this step is 18.6 kcal mol^{-1}, which is much lower than that of the elimination of tert-butanol in TBS without a catalyst.

[AlO_x] vs [SiO_x] insertion. In addition to the above mentioned attacking of TBS at the Al catalytic center, there is another possibility for the insertion of TBS at [SiO_x] site (Figure 11). One can find that at 500 K the TBS insertion at [AlO_x] is exergonic by 8.1 kca/mol, making the [AlO_x] attaching more favorable than the endoergic insertion at [SiO_x]. Due to Lewis acid charter of Al catalyst and strong Al-O interaction, TBS insertion at [AlO_x] is prior to insertion at [SiO_x]. The transition state of the elimination of tert-butanol by the insertion at [SiO_x] is also less stable than that by the insertion at [AlO_x]. Relative to the [AlO_x] insertion, the probability of TBS insertion at [SiO_x] site is rather small.

Elimination of isobutene. Subsequently, the new siloxane, **P1^C**, with two tert-butoxy groups further releases two isobutene ($H_2C = CH(CH_3)_2$) with the activation free energy of about 40 kcal/mol (Figure 11). The process of the elimination of isobutene is similar to that in silanol reaction (**B** reaction) through noncovalent catalysis of interfacial interactions. After all tert-butoxy groups change to –OH* via the elimination of isobutene, the propagation process increases a repeated unit, -$OSi(OH)_2$-. The geometry of the catalytic center [AlO_x] is almost same as that before TBS insertion (Figure 11). In the same way, the next TBS insertion into Al-O bond at [AlO_x] center leads to the propagation of the siloxane polymer on surface.

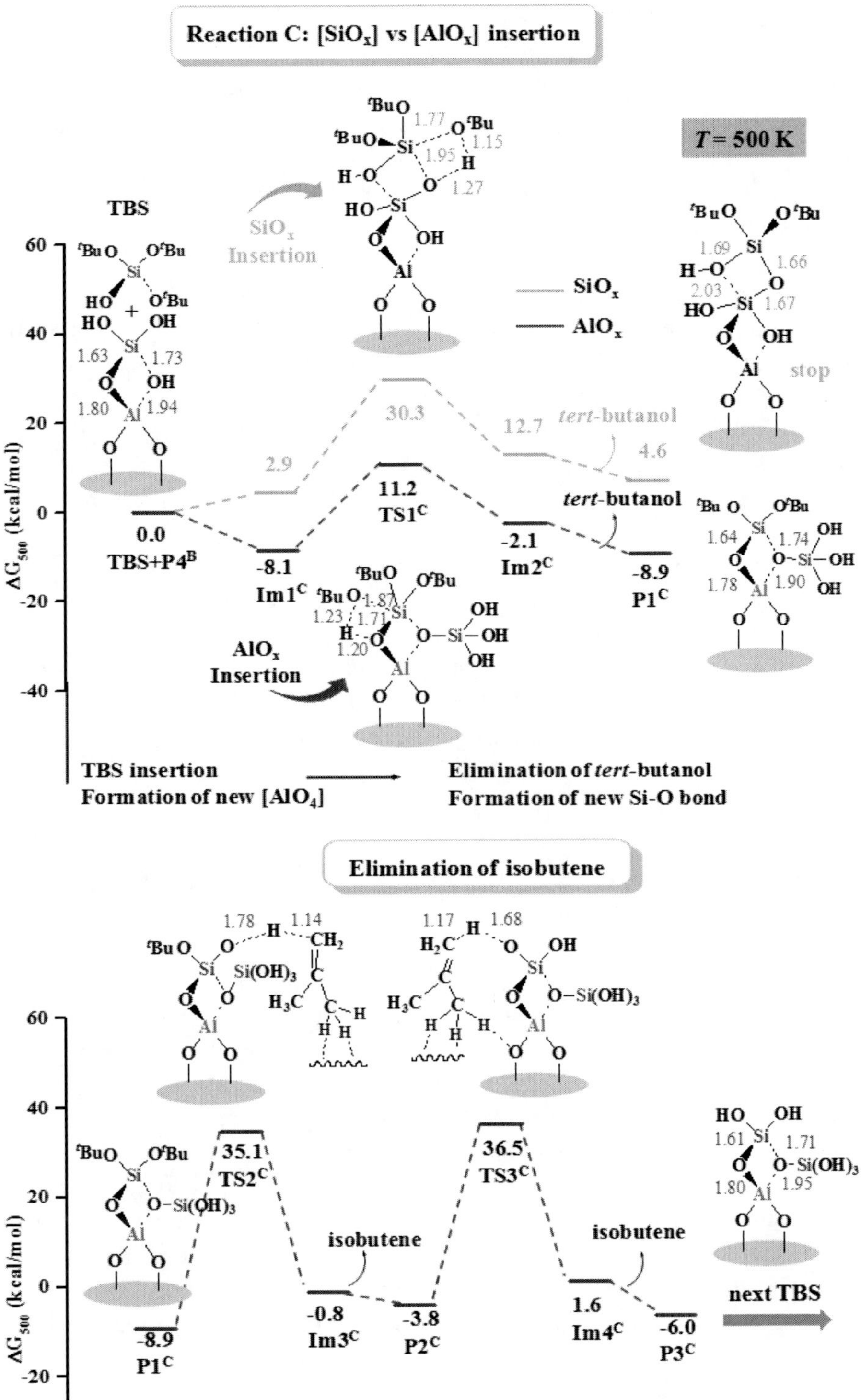

Reproduced from Reference [14] with permission from The Royal Society of Chemistry.

Figure 11. Gibbs free energy profile for the propagation reaction (**C**) of siloxane polymer on the hydroxylated SiO_2 surface.

(D) The Cross-Linking Reaction

After eliminating isobutene, the hydroxyl groups of siloxane polymer can react with other hydroxyl groups to yield H_2O. And then the new Si-O-Si bond is formed. These cross-linking reactions make the siloxane polymer layer gel and solidify to silica. At the same time, the cross-linking process limits the diffusion of TBS and eventually ceases insertion of TBS, which results in the self-limiting characteristic of the RALD reaction.

The high temperature may accelerate the cross-linking rate and inhibit the propagation rate of siloxane polymer. It is not surprising that in experiment RALD reactions occur at an optimum temperature range [11], which is higher than room temperature for isobutene elimination, but not too high. After cross-linking, sufficient hydroxyl groups are still left on the surface of silica. The RALD cycle may restart through the reaction with the next pluse of TMA provided that there are active hydroxyl groups on the surface.

To review the whole RALD cycle, including (**A**) TMA reaction, (**B**) silanol reaction, (**C**) propagation reaction, and (**D**) cross-linking reaction steps, we found that the introduction of TMA results in the formation of the Al catalytic site, which accelerates the propagation of the siloxane polymer.

The rate-determining step of whole RALD reaction is the elimination of isobutene of the tert-butoxy group, which can be catalyzed by the formation of Brønsted acid site of $[AlO_4]$ site and interfacial hydrogen bonding interactions between tert-butoxy group and the surface. The covalent catalysis and noncovalent catalysis work together in RALD reactions of Al_2O_3/SiO_2 nanolaminate. We expect that the synergistic catalysis of Lewis/Brønsted acid sites and interfacial interactions can be used in design and preparation of other silica nanolaminates, such as HfO_2/SiO_2, ZrO_2/SiO_2, and TiO_2/SiO_2.

MECHANISM OF PLASMA-ENHANCED ALD OF SiO_2

As an energy-enhanced ALD technique, plasma-enhanced atomic layer deposition (PE-ALD) has been developed to prepare high-quality SiO_2 thin films at low temperatures [41]. SiO_2 PE-ALD usually employs a Si precursor bearing an amino ligand and an O_2 plasma as the oxidant. Because it requires no catalyst and generates no corrosive by-product, SiO_2 PE-ALD has shown promising results and wide application [10, 41-43].

Recently, we have performed detailed density functional theory (DFT) calculations to investigate the mechanism of the reaction involving aminosilanes and O_2 plasma in SiO_2 PE-ALD [15]. It is demonstrated that the amino ligand can catalyze reaction of the aminosilane with surface hydroxyl groups (-OH). Such catalysis involving the amino group may be termed self-catalysis. The surface oxidation mechanism by the O_2 plasma is also discussed. The insights into the reaction mechanism of SiO_2 PE-ALD may be helpful to devise improved methods for preparing SiO_2 and other oxides, such as Al_2O_3, HfO_2, ZrO_2, and TiO_2.

(A) Aminosilane Half-Reaction

Similar to thermal ALD, SiO_2 PE-ALD comprises two half-reactions, involving the aminosilane (**A**) and the O_2 plasma (**B**) (Figure 12). Among Si precursors, aminosilanes with two amino ligands have been reported to be especially excellent candidates [10, 41-43]. In order to investigate the self-catalytic role of the amino ligand of the aminosilane, silane

(SiH_4) and BDMAS ($SiH_2[N(CH_3)_2]_2$) half-reactions (**A**) were selected for a comparative study.

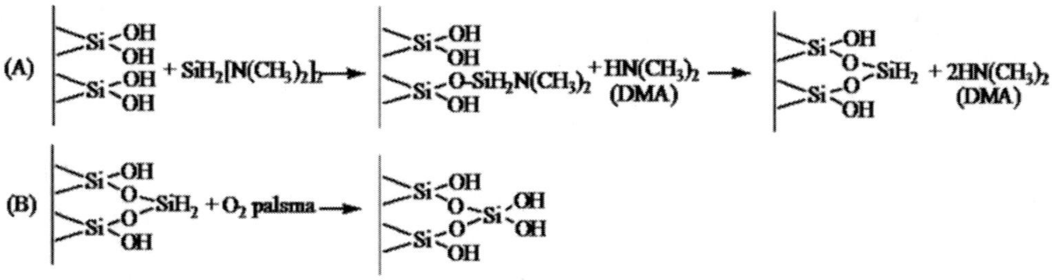

Reproduced from Reference [15] with permission from The Royal Society of Chemistry.

Figure 12. Aminosilane (A) and O_2 plasma (B) half-reactions in SiO_2 PE-ALD.

As shown in Figure 13, the activation free energy of the Si-O formation in the silane half-reaction amounts to 57.8 kcal mol^{-1}, indicating that the silane half-reaction is very difficult. When BDMAS is introduced, the activation free energy of the Si-O formation is dramatically lowered to about 11.0 kcal mol^{-1} and the aminosilane half-reaction proceeds very easily, which is consistent with the results of experiments at low temperatures, even room temperature [10, 41-43].

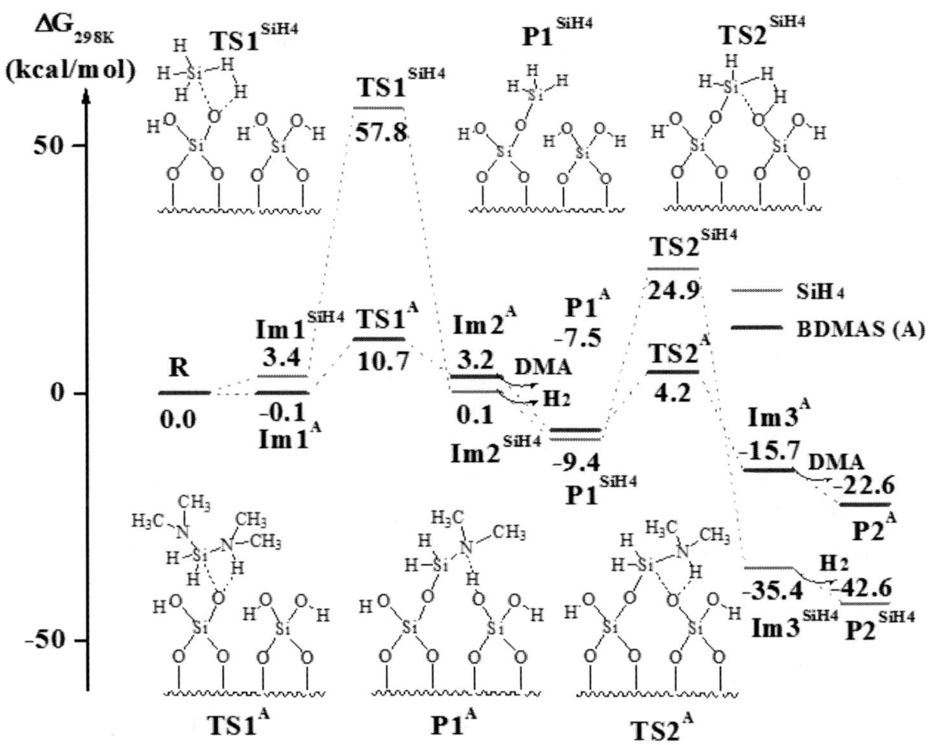

Reproduced from Reference [15] with permission from The Royal Society of Chemistry.

Figure 13. Gibbs free energy profiles for SiH_4 and BDMAS half-reactions.

In fact, the amino group of the aminosilane can remarkably catalyze the Si-O formation to generate SiO_2. We refer to the catalytic aminosilane reaction involving the amino ligand as self-catalysis. This self-catalysis by an amino ligand is similar to room temperature ALD (RT-ALD) of SiO_2 using Lewis-base catalysts, such as ammonia and pyridine, whereby the N atom can interact with hydroxyl (-OH) groups on the surface through strong N…H hydrogen bonding [8, 9, 13]. The self-catalysis of aminosilanes can also occur in the O_3-based ALD of SiO_2 [44, 45].

(B) O_2 Plasma Half-Reaction

Plasma is a high-energy aggregation state of matter, which includes a large number of electrons, ions, excited states of atoms, molecules, and free radicals. O_2 plasma species include triplet oxygen molecule ($^3O_2$), singlet oxygen molecules ($^1O_2$ or O_2*), triplet oxygen atoms (^3O), singlet oxygen atoms (^1O or O*), oxygen molecule cations (O^{2+}), oxygen cations (O^+), oxygen anions, (O^-), and electrons (e) (Figure 14). Among these species, $^3O_2$, $^1O_2$, ^3O, and ^1O are most abundant, which together constitute over 99.99% of the total [41, 46].

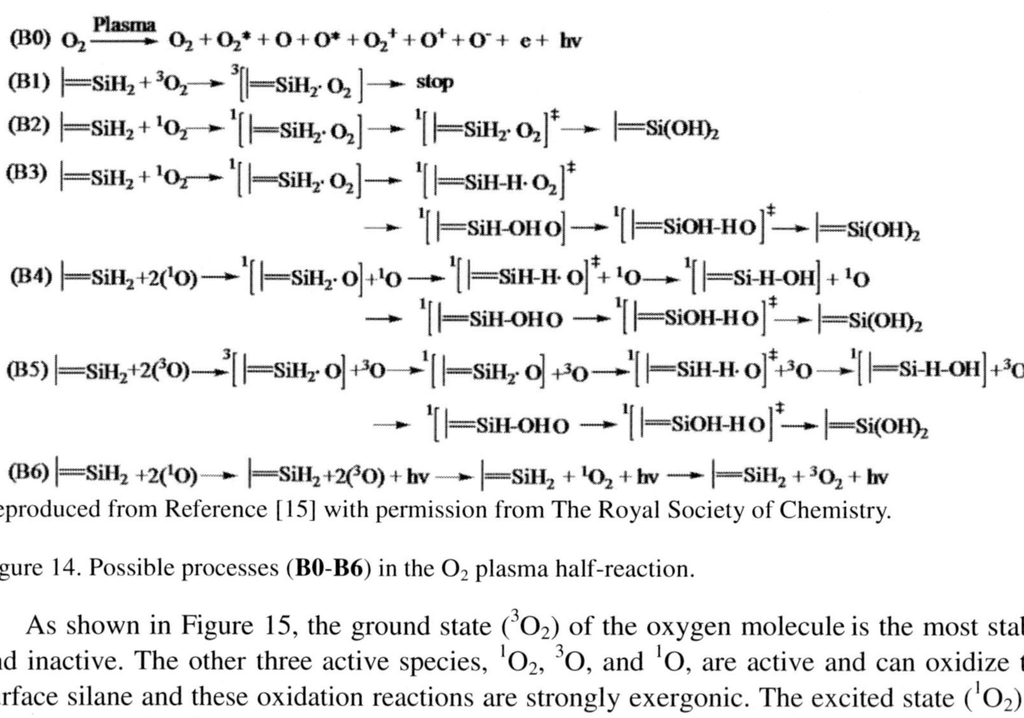

Reproduced from Reference [15] with permission from The Royal Society of Chemistry.

Figure 14. Possible processes (**B0-B6**) in the O_2 plasma half-reaction.

As shown in Figure 15, the ground state ($^3O_2$) of the oxygen molecule is the most stable and inactive. The other three active species, $^1O_2$, ^3O, and ^1O, are active and can oxidize the surface silane and these oxidation reactions are strongly exergonic. The excited state ($^1O_2$) of the oxygen molecule can go through one-step or stepwise pathway for oxidation of the silane. The overall oxidation reaction is strongly exergonic by 150.8 kcal mol^{-1}. In the one-step oxidation pathway, $^1O_2$ can frontally attacks two H atoms of the silane to derictly produce two -OH groups. In the stepwise pathway, $^1O_2$ firstly attacks one H atom of the silane from the side to generate the single -OH and then oxidize another Si-H bond to the product, -Si(OH)$_2$. Singlet oxygen atom (^1O) can oxidize the first Si-H bond on the surface to a single -OH product, -SiH-OH. Subsequently, another singlet oxygen atom (^1O) can insert the product -SiH-OH and form the singlet intermediate, which undergoes the second stepwise oxidation pathway of $^1O_2$. Ultimately, both Si-H bonds are oxidized to Si-OH groups. The triplet oxygen atom (^3O) interacts with the surface silane to form a triplet intermediate, which is converted to a singlet intermediate. The subsequent pathway is the same as the ^1O pathway. In the ^3O pathway, the oxidation can only proceed if the triplet complex is converted into a singlet intermediate. All active species, $^1O_2$, ^3O, and ^1O, can interact with the surface, which can result in the lose of their activity.

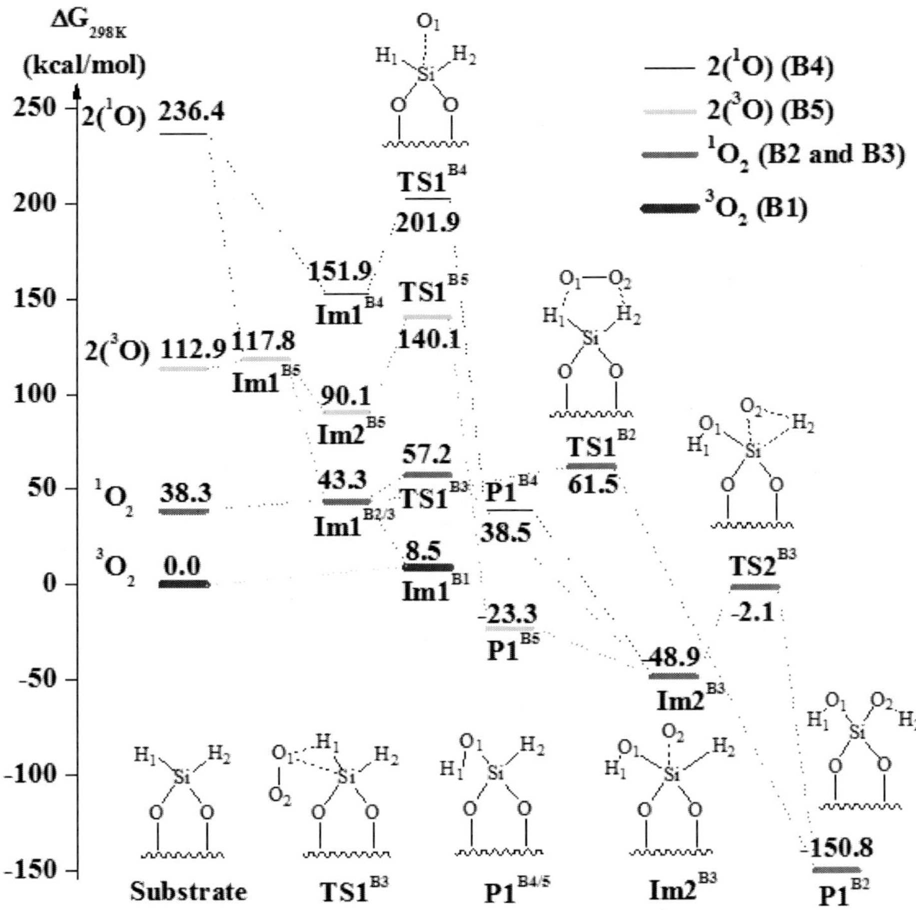

Reproduced from Reference [15] with permission from The Royal Society of Chemistry.

Figure 15. Gibbs free energy profiles for the O_2 plasma half-reaction.

When reviewing whole SiO_2 PE-ALD reactions, we find that the aminosilane plays a role of self-catalysis in Si-O formation through the relevant half-reaction involving its amino ligand. Among the various species in the O_2 plasma, $^1O_2$, ^1O, and ^3O are active and can strongly oxidize the surface silane groups. Oxidation of $^1O_2$ can undergo one-step or stepwise pathway. ^1O and ^3O can both oxidize two Si-H bonds. In the ^3O pathway, the triplet must be converted to a singlet and then follow the ^1O pathway.

We anticipate that the concept of self-catalysis of aminosilanes may be invoked in the design of more effective Si precursors for SiO_2, and the mechanism of strong surface oxidation by O_2 plasma may improve PE-ALD methodologies for the preparation of other oxides, such as Al_2O_3, HfO_2, ZrO_2, and TiO_2.

CONCLUSION

In summary, the reaction mechanisms of H_2O-based ALD, RT-ALD, RALD, and PE-ALD of SiO_2 have been investigated. In H_2O-based ALD of SiO_2, the $SiCl_4$ half-reaction may proceed through a stepwise pathway, first forming Si-O bond and then breaking Si-Cl/O-H bonds and forming H-Cl bond. H_2O half-reaction may undergo the hydrolysis and condensation processes, which can be catalyzed by H_2O molecules adsorbed on the surface.

In RT-ALD of SiO_2, Lewis base catalysts can dramatically reduce the activation energy of ALD reactions. The whole catalyzed reaction may undergo the multi-step pathways, including adsorption of precursor, axial addition, surface pseudorotation, axial elimination and desorption of byproduct.

In RALD of SiO_2, the synergistic catalysis of Lewis/Brønsted acid sites and interfacial interactions plays an important role in the RALD fabrication of silica nanolaminates. Lewis-acid catalyst can result in the formation of the catalytic site, which accelerates the propagation of the siloxane polymer. The rate-determining step of whole RALD is the elimination of isobutene of the tert-butoxy groups, which can be catalyzed by the Brønsted acid site of $[AlO_4]$ and the interfacial interactions, such as hydrogen bonding interactions between tert-butoxy groups and the surface.

In PE-ALD of SiO_2, the aminosilane plays a role of self-catalysis in Si-O formation through the relevant half-reaction. The three species of O_2 plasma, $^1O_2$, ^1O, and ^3O, can strongly oxidize surface silane groups to silanol (Si-OH). All these indicate that surface reactions of SiO_2 ALD are complex and may be exploited to improve ALD preparation of other oxides and their nanolaminates, such as Al_2O_3, HfO_2, TiO_2, Al_2O_3/SiO_2, and HfO_2/SiO_2.

ACKNOWLEDGMENTS

This work was supported by the State Key Program for Basic Research of China (2011CB922104), China Postdoctoral Science Foundation (2014M551556), and Zhejiang Provincial Natural Science Foundation of China (LY13B030005). Authors thank copyright permission of Reference [12-15] from Springer, American Chemical Society, and Royal Society of Chemistry, respectively. All these can be accessible via the Internet at http://dx.doi.org/10.1186/s11671-014-0714-1, http://dx.doi.org/10.1021/jp309974r, http://dx.doi.org/10.1039/c3nr02086j, and http://dx.doi.org/10.1039/c4cc08004a.

REFERENCES

[1] R. Doering and Y. Nishi, *Handbook of Semiconductor Manufacturing Technology*, 2nd ed., CRC Press, Boca Raton, FL (2007).

[2] M. Schulz, The end of the road for silicon? *Nature*, 399, 729-730 (1999).

[3] I. Kingon, J.-P. Maria and S. K. Streiffer, Alternative dielectrics to silicon dioxide for memory and logic devices, *Nature*, 406, 1032-1038 (2000).

[4] N. Pinna and M. Knez, *Atomic Layer Deposition of Nanostructured Materials*, Wiley-VCH, New York (2011).

[5] S. M. George, O. Sneh, A. C. Dillon, M. L. Wise, A. W. Ott, L. A. Okada and J. D. Way, Atomic layer controlled deposition of SiO_2 and Al_2O_3 using ABAB... binary reaction sequence chemistry, *App. Surf. Sci.*, 82/83, 460-467 (1994).

[6] O. Sneh, M. L. Wise, A. W. Ott, L. A. Okada and S. M. George, Atomic layer growth of SiO_2 on Si(100) using $SiCl_4$ and H_2O in a binary reaction sequence, *Surf. Sci.*, 334, 135-152 (1995).

[7] V. E. Drozd and V. P. Tolstoy, Study of the SiO_2 layer growth by the ML-ALE method, *App. Surf. Sci.*, 112, 269-272 (1997).

[8] J. W. Klaus, O. Sneh and S. M. George, Growth of SiO_2 at room temperature with the use of catalyzed sequential half reaction, *Science*, 278, 1934-1936 (1997).

[9] J. W. Klaus and S. M. George, Atomic layer deposition of SiO_2 at room temperature using NH_3-catalyzed sequential surface reactions, *Surf. Sci.*, 447, 81-90 (2000).

[10] A. Kobayashi, N. Tsuji, A. Fukazawa and N. Kobayashi, Temperature dependence of SiO_2 film growth with plasma-enhanced atomic layer deposition, *Thin Solid Films*, 520, 3994-3998 (2012).

[11] D. Hausmann, J. Becker, S. Wang and R. G. Gordon, Rapid vapor deposition of highly conformal silica nanolaminates, *Science*, 298, 402-406 (2002).

[12] G.-Y. Fang, L.-N. Xu, L.-G. Wang, Y.-Q. Cao, D. Wu and A.-D. Li, Stepwise mechanism and H_2O-assisted hydrolysis in atomic layer deposition of SiO_2 without a catalyst, *Nanoscale Res. Lett.*, 10, 68 (2015).

[13] G. Fang, S. Chen, A. Li and J. Ma, Surface pseudorotation in Lewis-Base-catalyzed atomic layer deposition of SiO_2: static transition state search and Born-Oppenheimer molecular dynamics simulation, *J. Phys. Chem. C*, 116, 26436-26448 (2012).

[14] G. Fang and J. Ma, Rapid atomic layer deposition of silica nanolaminates: synergistic catalysis of Lewis/Brønsted acid sites and interfacial interactions, *Nanoscale*, 5, 11856-11869 (2013).

[15] G.-Y. Fang, L.-N. Xu, Y.-Q. Cao, L.-G. Wang, D. Wu and A.-D. Li, Self-catalysis by aminosilanes and strong surface oxidation by O_2 plasma in plasma-enhanced atomic layer deposition of high-quality SiO_2, *Chem. Commun.*, 51, 1341-1344 (2015).

[16] J. K. Kang and C. B. Musgrave, Mechanism of atomic layer deposition of SiO_2 on the silicon (100)-2×1 surface using $SiCl_4$ and H_2O as precursors, *J. Appl. Phys.*, 91, 3408-3414 (2002).

[17] M. Ritala, K. Kukli, A. Rahtu, P. I. Räisänen, M. Leskelä, T. Sajavaara and J. Keinonen, Atomic layer deposition of oxide thin films with metal alkoxides as oxygen sources, *Science*, 288, 319-321 (2000).

[18] Y. Okamoto, Initial stage of the catalyzed growth of SiO_2 films on Si(001): an ab initio study, *J. Phys. Chem. B*, 103, 11074-11077 (1999).

[19] S. Chen, G. Fang, X. Qian, A. Li and J. Ma, Influence of alkalinity and steric hindrance of Lewis-Base catalysts on atomic layer deposition of SiO_2, *J. Phys. Chem. C*, 115, 23363-23373 (2011).

[20] R. S. Berry, Correlation of rates of intramolecular tunneling processes, with application to some Group V compounds, *J. Chem. Phys.*, 32, 933-938 (1960).

[21] C. Moberg, Stereomutation in trigonal-bipyramidal systems: a unified picture, *Angew. Chem. Int. Ed.*, 50, 10290-10292 (2011).

[22] E. P. A. Couzijn, J. C. Slootweg, A. W. Ehlers and K. Lammertsma, Stereomutation of pentavalent compounds: validating the Berry pseudorotation, redressing Ugi's turnstile

rotation, and revealing the two- and three-arm turnstiles, *J. Am. Chem. Soc.*, 132, 18127-181409 (2010).

[23] J. A. Deiters and R. R. Holmes, Pathways for nucleophilic substitution at silicon. a molecular orbital approach, *J. Am. Chem. Soc.*, 109, 1686-1692 (1987).

[24] J. A. Deiters and R. R. Holmes, Effect of entering and leaving groups on nucleophilic substitution reactions at silicon. A molecular orbital approach, *J. Am. Chem. Soc.*, 109, 1692-1696 (1987).

[25] R. R. Holmes, The stereochemistry of nucleophilic substitution of tetracoordinate silicon, *Chem. Rev.*, 90, 17-31 (1990).

[26] K. Mislow, Role of pseudorotation in the stereochemistry of nucleophilic displacement reactions, *Acc. Chem. Res.*, 3, 321-331 (1970).

[27] M. B. Tollefson, J. J. Li and P. Beak, The endocyclic restriction test: investigation of the geometries of nucleophilic substitution at phosphorus(III) and phosphorus(V), *J. Am. Chem. Soc.*, 118, 9052-9061 (1996).

[28] O. Legrand, J. M. Brunel and G. Buono, Stereospecific migration of P from N to C_{sp2}: ring-expansion reaction of chiral diazaphospholidine oxides, *Angew. Chem. Int. Ed.*, 38, 1479-1483 (1999).

[29] P. I. Dalko, *Enantioselective Organocatalysis: Reactions and Experimental Procedures*, WILEY-VCH Verlag GmbH and Co. KGaA, Weinheim (2007).

[30] E. P. Wasserman, I. Annis, L. J. Chopin, P. C. Price, J. L. Petersen and K. A. Abboud, Ethylene oxide polymerization catalyzed by aluminum complexes of sulfur-bridged polyphenols, *Macromolecules*, 38, 322-333 (2005).

[31] A. Motta, I. L. Fragalà and T. J. Marks, Links between single-site heterogeneous and homogeneous catalysis. DFT analysis of pathways for organozirconium catalyst chemisorptive activation and Olefin polymerization on γ-alumina, *J. Am. Chem. Soc.*, 130, 16533-16546 (2008).

[32] M. Linnolahti, J. R. Severn and T. A. Pakkanen, Formation of nanotubular methylaluminoxanes and the nature of the active species in single-site α-Olefin polymerization catalysis, *Angew. Chem., Int. Ed.*, 47, 9279-9283 (2008).

[33] A. Sokol, C. R. A. Catlow, J. M. Garcés and A. Kuperman, Computational investigation into the origins of Lewis acidity in zeolites, *Adv. Mater.*, 12, 1801-1805 (2000).

[34] S. Malola, S. Svelle, F. L. Bleken and O. Swang, Detailed reaction paths for zeolite dealumination and desilication from density functional calculations, *Angew. Chem., Int. Ed.*, 51, 652-655 (2012).

[35] J. To, A. A. Sokol, S. A. French, N. Kaltsoyannis and C. R. A. Catlow, Hole localization in $[AlO_4]^0$ defects in silica materials, *J. Chem. Phys.*, 122, 144704 (2005).

[36] W. Wang and M. Hunger, Reactivity of surface alkoxy species on acidic zeolite catalysts, *Acc. Chem. Res.*, 41, 895-904 (2008).

[37] S. L. Wegener, T. J. Marks and P. C. Stair, Design strategies for the molecular level synthesis of supported catalysts, *Acc. Chem. Res.*, 45, 206-214 (2012).

[38] J. Čejka, A. Corma and S. Zones, *Zeolites and Catalysis, Synthesis, Reactions and Applications*, WILEY-VCH Verlag GmbH and Co. KGaA, Weinheim (2010).

[39] F. Weinhold and C. R. Landis, *Valency and Bonding: A Natural Bond Orbital Donor-Acceptor Perspective*, Cambridge University Press, Cambridge (2005).

[40] F. Weinhold and C. R. Landis, *Discovering Chemistry with Natural Bond Orbitals*, John Wiley and Sons, Inc., Hoboken, New Jersey (2012).

[41] H. B. Profijt, S. E. Potts, M. C. M. van de Sanden and W. M. M. Kessels, Plasma-assisted atomic layer deposition: basics, opportunities, and challenges, *J. Vac. Sci. Technol., A*, 29, 050801 (2011).

[42] S.-J. Won, S. Suh, M. S. Huh and H. J. Kim, High-quality low-temperature silicon oxide by plasma-enhanced atomic layer deposition using a metal-organic silicon precursor and oxygen radical, *IEEE Electron Device Lett.*, 31, 857-859 (2010).

[43] Y. Lu, A. Kobayashi, H. Kondo, K. Ishikawa, M. Sekine and M. Hori, Chemical reactions during plasma-enhanced atomic layer deposition of SiO_2 films employing aminosilane and O_2/Ar plasma at 50°C, *Jpn. J. Appl. Phys.*, 53, 010305 (2014).

[44] B. Han, Q. Zhang, J. Wu, B. Han, E. J. Karwacki, A. Derecskei, M. Xiao, X. Lei, M. L. O'Neill and H. Cheng, On the mechanisms of SiO_2 thin-film growth by the full atomic layer deposition process using bis(t-butylamino)silane on the hydroxylated SiO_2(001) surface, *J. Phys. Chem. C*, 116, 947-952 (2012).

[45] L. Huang, B. Han, B. Han, A. Derecskei-Kovacs, M. Xiao, X. Lei, M. L. O'Neill, R. M. Pearlstein, H. Chandra and H. Cheng, First-principles study of a full cycle of atomic layer deposition of SiO2 thin films with di(sec-butylamino)silane and ozone, *J. Phys. Chem. C*, 117, 19454-19463 (2013).

[46] M. A. Lieberman and A. J. Lichtenberg, *Principles of Plasma Discharges and Materials Processing*, 2nd ed., John Wiley and Sons, Hoboken, NJ (2005).

In: Atomic Layer Deposition (ALD)
Editor: Jeannie Valdez
ISBN: 978-1-63483-869-6

Chapter 5

PLATINUM NANO-THIN FILM FOR PLASMONIC PHOTOCATALYTIC REACTION

Hung Ji Huang** and Bo-Heng Liu
Instrument Technology Research Center,
National Applied Research Laboratories, Taiwan

ABSTRACT

Plasma-enhanced atomic layer deposition (PEALD) is a power tool that can fabricate high-quality single-crystal metal or dielectric thin films. In this chapter, the low-temperature fabrication process of single-crystal platinum (Pt) nano-thin films using PEALD was conducted without the use of a predeposited seed layer. The Pt thin metal films deposited on Si and glass substrates exhibited high conductivities and thermal stabilities resembling those of the bulk material in crystal form. The Pt thin films with good thermal and chemical stability are good candidates for use in photocatalytic reactions. The deposited Pt thin films by PEALD were chemically stable under light illumination (50 W halogen lamp) and could serve as catalysts under strongly alkaline conditions (pH = 12) during the long-term oxidization of ammonium ions. The deposited Pt thin film served as a one-dimensional quantum well for the transformation of plasmon energy into chemical energy. The energy transformation in metal structures with confinement at the nano scale provides an interesting physical discussion of quantum electrodynamics.

1. INTRODUCTION

The atomic layer deposition (ALD) method is an important method for the deposition of ultra thin films in semiconductor fabrication processes. In ALD methods, crystal thin films are grown one atomic layer at a time [1, 2]. The precursor gases and vapors are alternatively injected on the substrate surface, leading to subsequent chemisorption and surface reaction of the precursors [3]. A self-limiting growth mechanism leads to the growth of conformal thin

* E-mail address: hjhuang@narlabs.org.tw, hjhuang@itrc.narl.org.tw.

films over large areas with an accurately defined thickness in deposition processes [1-5]. Various dielectric materials have been deposited using these technologies, including Al_2O_3 [4, 6-9], polycrystalline luminescent ZnS:Mn [10], TiO_2 [5, 8, 10, 11], TaN_x [12], ZrO_2 [13], $HfO_2Ta_2O_5$ [4], PtO_x [14], and ZnO [15, 16]. Conventional thermal ALD systems were used to deposit materials (HfO_2, Al_2O_3, TiO_2, etc.) with high dielectric constants [9, 17, 18]. Many types of metals have also been deposited using the ALD method, including Ru [17, 19-21], Pt [14, 22-24], W [25], and Cu [26].

Pt thin films exhibit high chemical stability, high thermal stability, and excellent electrical conductivity, which makes them potentially useful in numerous applications. In ALD deposition, Pt thin films were deposited onto TaN_x using a thermal ALD method with a seed/barrier stack and exhibited good step coverage [22]. In thermal ALD, reactions are confined to a limited number of reactive sites on the SiO_2 substrate, thereby forming isolated Pt nanoparticles (NPs). The uniform Pt NPs were deposited at a high substrate temperature of 300 °C using the precursor (methylcyclopentadienyl) trimethylplatinum ($MeCpPtMe_3$) and oxygen [22, 23, 27]. The Pt NPs became larger as the cycle number of ALD was increased, and the coalesced nanocrystals finally formed a rough film. At a low substrate temperature of 130 °C, only PtO_x films were obtained from the combination of $Pt(acac)_2$ with O_3. Thus, it is not easy to fabricate thin, smooth Pt layers of homogeneous structures.

The inductively coupled PEALD method is a useful ALD technology that incorporates a powerful inductively coupled plasma generator. In semiconductor fabrication, inductively coupled plasma is essential in dry etching to construct specific structures on InP, Si, and others. Inductively coupled plasma has also been applied to mass spectrometry and to the fabrication of carbon nanowalls, graphene, GaN nanorods, etc. Recently, the PEALD method has been used to fabricate metal nitrides and metal films with high conductivity. This deposition process was developed for microelectronic applications such as large-scale film deposition [12, 19-21, 28-30] or plasmonic oxidization of ammonium [31, 32].

In this chapter, we present how to fabricate Pt thin films using the PEALD method used in our research [29, 31, 32]. The Pt metal thin films deposited using the PEALD method are thin, smooth, and exhibit high conductivity similar to that of the bulk material. Pt is stable in chemical catalytic reactions and exhibits a high optical plasmonic response [33-37]. Deposited Pt thin films also exhibit high thermal stability and good durability under harsh environments for the chemical decomposition of ammonium (NH_4^+). The deposited Pt thin film served as an one-dimensional quantum well for the transformation of plasmon energy to chemical energy. The energy transformation in metal structures with confinement on the nanoscale enables an interesting physical discussion of quantum electrodynamics.

2. Low-Temperature PEALD

Thin, homogeneous Pt thin films were deposited using high-energy reductive plasma-enhanced deposition in a home-built PEALD system without any predeposited Pt seed layer [29, 31, 32]. A cycled process [29] of alternate pulsing the precursor gases and vapors onto the substrate surface leads to subsequent chemisorption and surface reaction of the precursors as shown in Figure 1. First, the vertical flow method was used to introduce the ALD reactants of the precursor vapor into the chamber to react on the substrate. Second, the purge gas (N_2)

was introduced to remove volatile byproducts from the process chamber. Third, the O_2 plasma source was generated at 13.56 MHz in a radio frequency matching box. Plasma is composed of charge carriers of ionized molecules or atoms that can be controlled through the application of electromagnetic fields. The growth mechanism of thin Pt thin films and their quality depend on the pressure and plasma exposure time [14, 19]. To avoid chemical vapor deposition, we used an automated shutter in the PEALD process to independently control the O_2 plasma exposure time and the Pt precursor injection-pulse time [29]. Fourth, the purge gas (N_2) was introduced to remove volatile byproducts from the process chamber. In this ALD method, the complete processing cycle leads to a self-limiting surface and provides excellent control of the atomic layer thickness and good step coverage in ultrahigh-aspect-ratio trench structures. To increase the thickness of the thin film, we repeated the reaction cycles a predetermined number of times. The self-limiting reactions ensure that the film thickness increases cycle by cycle. The Pt thin film growth rate was approximately 0.4 Å per cycle.

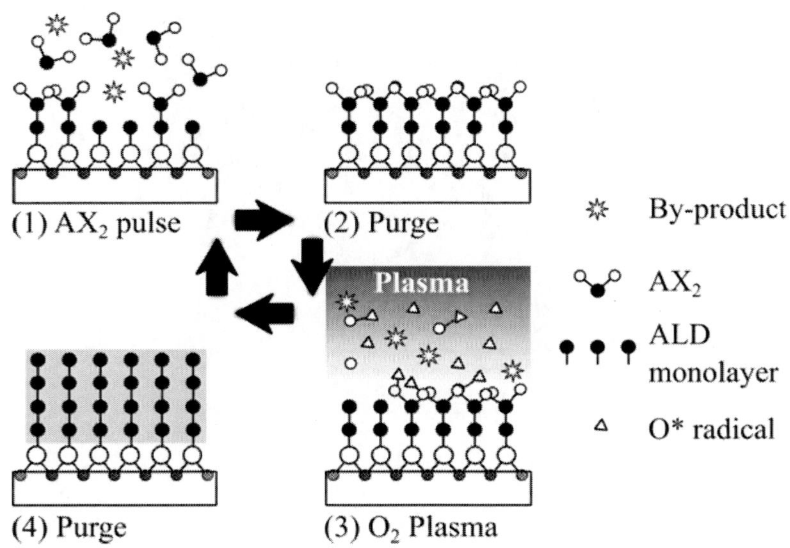

Figure 1. Schematics of the reaction process in PEALD.

A quartz tube connected to an inductively coupled plasma source served as the plasma chamber of the home-built PEALD system. The induced plasma gases were allowed to enter the deposition chamber only during controlled periods and were confined by a remote-controlled built-in shutter. The $MeCpPtMe_3$ (Sigma–Aldrich, purity 98%) precursor vapor was generated at 70 °C and mixed with Ar/O_2 gas under plasma when injected at predetermined pulse times. The additional Ar/O_2 plasma exposure time was 3 s. The target substrate was maintained at various temperatures to achieve an optimized film quality. At the end of each cycle, the purge gas (N_2, 99.9995%) was introduced in 5 s pulses to remove the exhaust gas in the process. Herein, the Pt layers were deposited onto a Si (100) substrate with a native oxide layer as well as on glass substrates, as shown in Figure 2.

In this chapter, the platinum thin films were fabricated using the home-built PEALD system at low substrate temperatures. The high-energy reductive plasma was demonstrated to effectively induce the growth of crystallized Pt thin films using the PEALD method. The Pt seed layer was not pre-made for any of the fabrication processes.

A diffractometer with an 18 kW rotating anode X-ray source (Rigaku TTRAX III) was used to examine the crystal structure of the deposited layers based on low-angle (< 2°) X-ray diffraction (XRD) measurements. X-ray reflectivity measurements were also used to determine the deposited thickness and density of the Pt thin films. The X-ray photoelectron spectroscopy (XPS) spectra of the deposited thin film were analyzed using an ESCA PHI 1600 spectrometer with a Schottky thermal field emitter (Perkin Elmer PHI 670). The resistivity was measured using a Hall-effect measurement system (Accent/HL 5500PC). The film morphology and accurate thickness were determined using transmission electron microscopy (TEM, JEOL JEM-2100F, accelerating voltage: 160–200 kV). The surface roughness was determined using atomic force microscopy (AFM, Veeco Dimension 3100, processed in tapping mode).

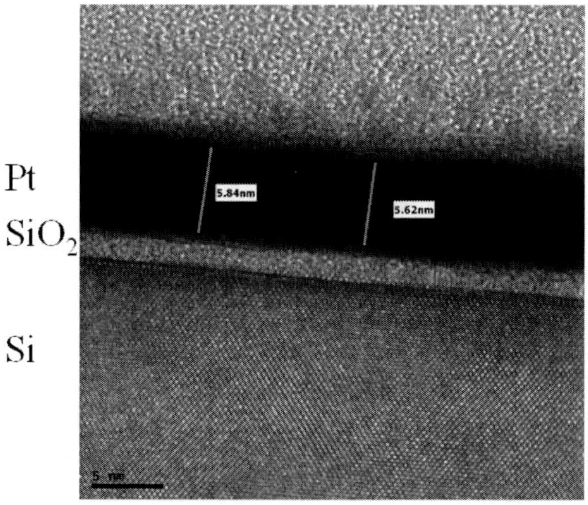

Figure 2. TEM image of the deposited Pt ultra thin film.

At substrate temperatures of 175 °C and 200 °C, the deposited Pt thin film exhibited a doublet peak representing bonding energies of 74.3 eV (4f5/2) and 71.1 eV (4f7/2) in XPS measurements calibrated with respect to carbon 1s (285.5 eV). These peaks resembled those reported elsewhere for Pt [38]. In the XRD measurements, the Pt thin films exhibited a high-intensity preferred-orientation (111) peak [fcc structure], similar to that observed in other studies [14, 23]. A polycrystalline Pt complex layer was formed when the substrate temperature was set at 100 °C. A smooth Pt thin film was regularly obtained at a substrate temperature of 200 °C. The resulting film can be considered a quasi single crystal.

The roughness of the 8-nm-thick Pt thin film on the Si substrate was measured by AFM to be 0.368 nm. The measured density of the deposited Pt thin films was 20.76 g/cm^3. The deposited Pt layer was a continuous and smooth crystal thin film with ultra small thickness that can be considered a quantum well of electrons and plasmons.

The resistivities of the 8-nm-thick Pt thin films on Si substrates with native oxide prepared at substrate temperatures of 175 °C and 200 °C were 20 and 16 μΩ cm, respectively. For comparison, Pt thin films with similar resistivity required substrate temperatures exceeding 300 °C when deposited using the thermal ALD method [23, 29]. The Pt thin metal films deposited by PEALD at 200 °C exhibited high conductivities (<12 μΩ cm for films with a thickness greater than 8 nm, resembling that of the bulk material). Low resistivity means a high concentration of free electrons in the metal layer and great optical response to external electromagnetic waves, which is important for plasmonic-related applications.

The PEALD method can be used to deposit nanometer-sized Pt thin film on a large surface area with substantial variations in surface structure. The homogeneous and continuous thin film can be deposited on surfaces of many trench bottoms and walls [22, 29] with good step coverage. A 25.49-nm-thick Pt layer was deposited to cover a 170-nm-wide SiO_2 trench structure with an aspect ratio of 3.5:1. The low-temperature PEALD process was demonstrated to yield good step coverage [29]. This method provides the possibility of fabricating components with feature sizes of less than 20 nm. For advanced applications, the deposition of nanometer-sized thin films using the PEALD method also provides the possibility of conductive layer filling at a very high aspect ratio by a hole. The low-substrate-temperature PEALD process also provides a potential coating method with the possibility of causing less damage to substrates with premade structures or materials during fabrication.

The deposited Pt thin films exhibited good thermal and chemical stability. The PEALD-deposited Pt thin films were chemically stable under high-temperature light illumination and could serve as catalysts under strongly alkaline conditions (pH = 12) during the long-term oxidization of ammonium ions. The stability of PEALD-coated Pt thin films also provides possibilities for advanced applications. In this chapter, the Pt thin film was used as a photocatalyst to enhance the decomposition of the endothermic ammonium oxidization process. The ultra small thickness causes the Pt thin film to act as a one-dimensional quantum well for plasmonic energy transformations.

Plasmons [39-54] are quantized collective motion of free electrons in metal that generate strong localized electromagnetic fields and localized hot spots for plasmonic heating [55] and many applications. The collective motion of free electrons responds to incoming external light and produces homogeneous plasmons on a metal thin film [44, 56] and enhances the chemical reaction. This process is called a plasmonic photocatalytic reaction and is very important for wet chemistry for green environmental technology. However, the used metal must be stable in a chemical reaction or catalytic reaction under harsh conditions. Au, Ag, Fe, Zn, Sb, Cu, and Al exhibit strong optical plasmonic responses. Au is very stable but is only used as a catalyst with a size of a few nanometers in some special chemical reactions. Ag, Fe, Zn, Sb, Cu, Al, and other metals exhibit high catalytic efficiencies and optical plasmonic responses. However, these metals are not very stable and tend to oxidize. Pt is stable in chemical catalytic reactions [31-37, 55] and exhibits a high optical plasmonic response. However, the deposition of a Pt thin film on a reaction disk to be used in a spinning-disk reactor (SDR) is very difficult and easily strips off in chemical processes. The PEALD method can well control the final deposited thickness of the film and produces samples with good film adhesion. The properties of the deposited Pt thin films resemble those of bulk Pt, which makes them good candidates for plasmonic photocatalysts. The results indicate that the Pt thin films were deposited with precisely controlled thickness. The low sheet resistivity of the deposited thin film indicates a high concentration of free electrons for inducing a sufficient optical plasmonic response. Our home-built PEALD system allows the large-area coatings of Pt thin films on the glass disks (12 cm in diameter) for use in plasmonic photocatalytic ammonium oxidization [31, 32].

3. CATALYTIC AMMONIUM OXIDIZATION

The ammonia or ammonium in water generated from organic pollutants is becoming a severe problem [57-61]. In natural environments, the NH_4^+ produced by organic materials can be reduced in the nitrogen cycle [62, 63]. Biological nitrification and denitrification processes typically occur in aquariums. Nitrosomonas bacteria, nitrospira bacteria, algae, and water plants help complete the nitrogen cycle. However, the nitrogen cycle may cease when the NH_4^+ concentration, temperature, or pH of the water suddenly changes.

Previous studies have presented methods for artificial nitrogen cycle processes using Pt as a catalyst [35, 37, 64, 65]. A high system temperature was necessary to yield a high processing efficiency in the ammonium oxidization using the typical wet air oxidization (WAO) method. A well-sealed reaction chamber was compulsory as the system temperature must be higher than the boiling point of water. An additional shield or protection may also be required to ensure the safety of the entire system. In such artificial processes, the ammonia removal efficiency was largely affected by the initial pH and ammonia concentration [37].

The high-temperature and high-pressure environment in the WAO method is introduced to increase the mass transfer rate and processing efficiency. However, the WAO method requires large power consumption to maintain these conditions, which is not convenient as the treated water must be cooled down before being discharged.

The ammonium ions in water were treated with the home-built SDR, see Figure 3, for high processing efficiency. The SDR can especially enhance the reactions needed for high mass transfer rates or chemical contact rates in air or liquid. In the photocatalytic/catalytic reaction, the final desorption of products is a typical bottleneck of chemical reactions. The strong affinity of the products to bind to the photocatalyst/catalyst strongly suppresses the chemical processing efficiency. A servomotor rotated the glass reaction disk (B270, diameter = 120 mm) in the reactor at a constant speed of 150 rpm in the experiments in this chapter. The rotation of the reaction disk generates many micro-vortices when the target solution flows over. Therefore, a SDR can have a high mass transfer rate to strip the products from the photocatalyst/catalyst and deliver the reactants. The photocatalytic/catalytic reaction can be substantially enhanced. An SDR exhibits a high processing efficiency and low power consumption and occupies a small space for mass treatment [29, 33, 66-72].

Furthermore, a light source of a halogen lamp with an aluminum bow reflector illuminated an area of approximately 2 cm × 5 cm on the reaction disk and supplied extra energy to enhance the chemical reactions. The chemical stability of the deposited Pt thin film is critical to its use in SDR or other chemical reactors under harsh processing conditions.

In this chapter, serious of experiments presented the use of deposited Pt thin film by PEALD in SDR. Before initiating the light-illumination experiments, the halogen lamp was turned on for 3 min to warm both the lamp and reaction disk. In the reference experiments without light illumination, the experiments with pH adjustment started 3 min after the addition of sodium hydroxide (NaOH) solution. The processing time with SDR was 30 min. The sample solution was circulated using a peristaltic pump (BT 102S, Yeong Shin Co., Ltd., Taiwan) at an average flow rate of 89 mL/min. The sample solution was dispensed onto the reaction disk in rotation, collected in a glass container (Schott Duran, Germany), and circulated again. A 100-mL sample solution of dissolved NH_4Cl (2.3 mM, Mallinckrodt Baker, Inc., USA) was prepared. The test solution was added with NaOH solution to adjust the pH value and supplied OH^- ions for the oxidization of NH_4^+. The concentrations of the dissolved chemicals in the test solution (NH_4^+, NO_2^-, and NO_3^-) were measured before and after the process using a digital water analyzer (Lambda-9000, Kyoritsu Chemical-Check Lab. Corp., Japan). The processes used in the experiments were kept consistent.

The room-temperature oxidation of NH_4^+ was processed with OH^- under alkaline conditions [31] to reach pH 11 and pH 12, Table 1. Reference experiments were performed without adding OH^- to represent the process of "no" chemical reaction. Interestingly, the NH_4^+ oxidization occurred in the SDR under ambient conditions without the high-temperature and high-pressure environment used in the typical WAO method [31]. The SDR proved to present the high processing efficiency in ambient conditions.

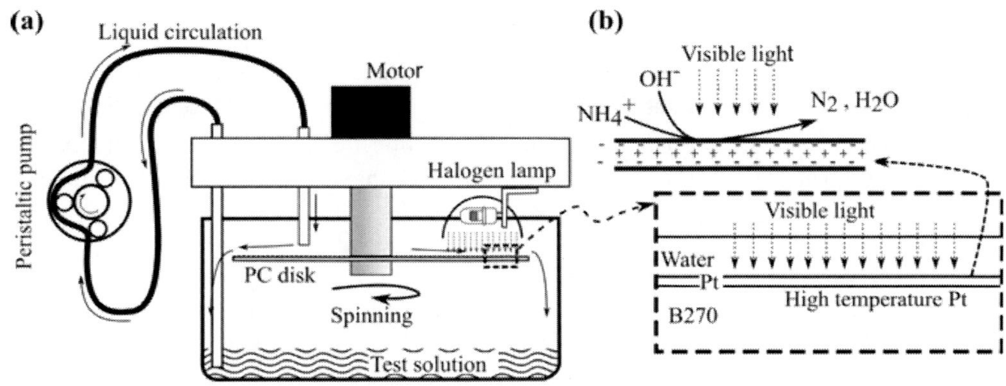

Figure 3. Schematics of plasmonic ammonium oxidization.

In addition, illumination from a halogen lamp can enhance the processing efficiency by increasing the temperature of water in experiments with a bare glass reaction disk. In the experiments with Pt thin films, the halogen lamp illumination induced a hot area on the reaction disk to support the additional plasmonic photocatalytic reaction. The halogen lamp enhanced the processing efficiency of the NH_4^+ endothermic oxidization. The stability and optical response of Pt thin films are important for the plasmonic photocatalytic reaction. The PEALD-coated Pt thin film was beneficial with its high stability under high temperature and large variation in temperature in the long-term processes of NH_4^+ under strongly alkaline conditions (pH = 12) [31, 32]. The concentration of dissolved ammonium in the 100-mL solution decreased from 2.3 (originally) to 0.41 mM at the end of the experiment. In this chemical reaction, OH^- was used to oxidize NH_4^+. Therefore, the pH decreased from 12 to 11.51. These results indicate that the PEALD-coated Pt thin film is useful as a active catalyst in water treatment and pollution reduction.

Table 1. Oxidization of NH_4^+ in water. (data from ref [31])

	Pt_No OH^-	Pt_pH11	ΔT_1_pH11	Pt_pH12	ΔT_1_pH12
Final NH_4^+	2.34	0.39		0.41	
Final temperature	29.15	28.93	−0.22	28.55	−0.60
	B270_No OH^-	B270_pH11	ΔT_1_pH11	B270_pH12	ΔT_1_pH12
Final NH_4^+	2.40	0.67		0.61	
Final temperature	27.87	27.73	−0.45	28.10	−0.55

The use of an SDR is a proficient method for reducing highly concentrated pollutants in air/water. It can be incorporated into a versatile water treatment system to achieve processing results that are optimized for a specific environment. The cascadable setup of the SDR [73] can provide high processing efficiency for composite chemical/organic pollutants in water. Various photocatalysts, e.g., TiO_2 [68, 73], ZnO [69], and Pt [31, 32], have been demonstrated to be highly efficient and effective photocatalysts in SDRs. Therefore, the incorporation of SDRs is an important green environmental technology for modern chemical engineering.

4. Energy Transformation

The light illumination on the Pt nano-thin film induced additional effects of energy compensation in the NH_4^+ oxidization. The relative absorbed heat (Δq) was calculated as a function of the material properties [74], and the results are presented in Table 2, which provide further information on the effects of plasmonic enhancement mediated by the Pt thin film.

$$\Delta q = \Delta T_i \sum_j m_j C_{p\_j} \tag{1}$$

The experimental "plasmonic endothermic reaction" and "endothermic reaction" results were obtained with a reaction disk of "B270 substrate coated with Pt thin film" and a "raw disk," respectively. In the calculations, $\Delta q < 0$ for the solutions in (plasmonic photocatalytic) ammonium oxidizations. Furthermore, the difference in absorbed heat was greater than zero, i.e., $\Delta q_{ppc} - \Delta q > 0$. The plasmonic photocatalytic endothermic reaction requires less thermal energy than the normal endothermic chemical oxidation of NH_4^+.

Table 2. Chemical energy for thermal and plasmonic NH_4^+ oxidation [32]

	PPC endothermic reaction		Endothermic reaction		
Initial pH	ΔT_1 (°C)	Δq_{ppc} (kJ/mol)	ΔT_2 (°C)	Δq (kJ/mol)	$\Delta q_{ppc} - \Delta q$
11	−0.22	−0.874	−0.45	−2.865	1.992
12	−0.60	−2.386	−0.55	−3.444	1.058

From the experimental data, light absorption is important in the ammonium oxidization by SDR. The classical approach based on Snell's law and light attenuation in the material can provide a simple picture. The vertically illuminating light encountered multiple steps of transmission, absorption, and reflection. Part of the light energy was absorbed by the layers of water, Pt, and glass. The electromagnetic wave is gradually eliminated with increasing propagating distance. The intensity of light, I, is a function of the square of the electromagnetic wave:

$$I(\omega) = |E(z,t)|^2 = E_0^2 e^{-4\pi n_{img} z/\lambda_0} = I_0 e^{-\alpha z} \tag{2}$$

where $I_0 = |E_0|^2$. The attenuation constant is

$$\alpha = \frac{4\pi n_{img}}{\lambda_0} = \frac{4\pi Im(\sqrt{\epsilon\mu})}{\lambda_0} \quad (3)$$

The light intensity gradually decreases when it propagates inside a material with a non-zero imaginary part of the refractive index. The light absorption increases with the long propagating distance in a specific layer after numerous reflections. The incident light propagates to the bottom surface of each layer. The light energy absorbed by each layer can be evaluated based on the refractive index. The imaginary part of the refractive index, Figure 4, exhibits the strong absorption of light. The water and glass layers largely absorb light in the infrared band in the experiments, which increases the system temperature.

In a heterogeneous catalytic reaction, the variation of ambient temperatures or the catalyst affects the reaction rate. The Arrhenius equation gives the dependence of the chemical reaction rate constant k on the absolute temperature T:

$$k = Aexp\left(\frac{-E_a}{k_B T}\right) \quad (4)$$

where E_a is the activation energy of various chemicals, A is a constant, and k_B is the Boltzmann constant or gas constant. The increases of the system temperature T enhance the reaction rate, k. However, the differences in the final temperature for the four different experiment groups were small, see Table 1. The Arrhenius equation does not provide further understanding of this issue.

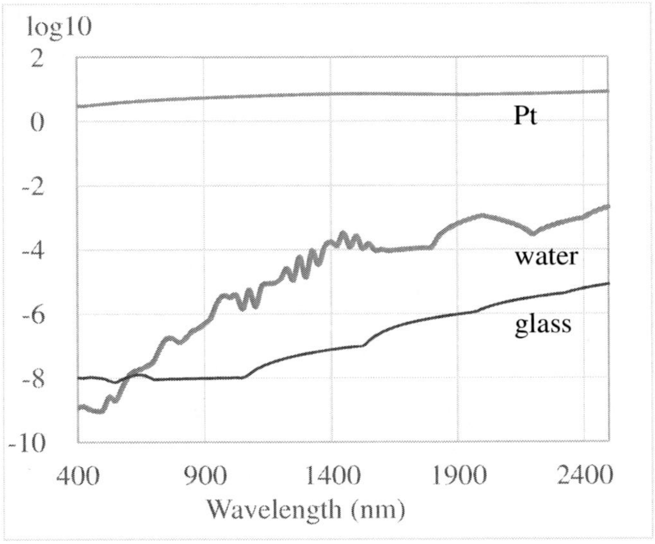

Figure 4. Imaginary part of the refractive index for various materials [75].

In the experiments with Pt thin films, the vertical illuminated light further induces the appearance of plasmons, the collective motion of free electrons, on the Pt layer. The induced plasmons can radiatively decay into light again. The induced plasmon can non-radiatively decay into heat through plasmon–phonon interactions. The generated heat from plasmon or light illumination can also increase the temperature of the system, similar to the phenomenon observed in layers of glass and water. This temperature increase can further enhance the chemical reaction through the model of Arrhenius equation. However, the energy difference in the chemical reaction remains unresolved.

The Pt nano-thin film is the catalyst for chemical reactions in SDR and acts as a one-dimensional quantum well for plasmons and electrons, which can have different effects on the photocatalytic reaction. Recently, many research works have suggested that quantum hot charges (QHC) can further trigger some chemical reactions that do not easily occur in typical ambient conditions [31, 32, 76-91]. The low sheet resistivity of Pt results in a high concentration of free electrons, which can be modulated to oscillate with external light illumination and induce a high optical plasmonic response and generate plasmons. Some interesting studies discovered that the induced plasmon can also decay to deliver energy to induce a single high-energy QHC [84, 85]. The one-dimensional quantum well of the Pt thin film can mediate the plasmon–QHC decay process. The high-energy QHC on the surface of metal can directly deliver energy to enhance the chemical reaction and compensate the energy needed in the endothermic ammonium oxidization.

Furthermore, in the SDR, the absorbed light transforms into plasmonic resonance energy on the platinum thin film and forms a high-temperature hot layer, which can act as another path of energy compensation through the generated hot electrons [91].

Electrons are fermions with 1/2 spin. The Fermi–Dirac distribution,

$$f(E) = \frac{1}{exp\left(\frac{E-E_F}{k_B T}\right)+1} \tag{5}$$

describes the probability of a fermion staying at a state with energy E under the restriction of the Pauli exclusion principle. k_B is the Boltzmann constant, T is the absolute temperature of the system, and the Fermi energy (E_F) is the highest occupied energy state of fermions in a stationary system. The illuminated light increases the temperature of the Pt thin film. The population of electrons with energy higher than E_F largely increase and become high-energy hot electrons. The energy of plasmons transforms into heat and induces a high concentration of hot electrons on the surface of the Pt thin film. The hot electrons can also deliver energy to compensate for the chemical energy required by the endothermic reaction. The homogeneous Pt thin film is a one-dimensional quantum well that also serves as a medium to transfer light energy through the generation of hot electrons.

Interestingly, both model of the hot electrons in the Fermi–Dirac distribution and the QHCs generated from the non-radiative decay of plasmons can enhance the plasmonic photocatalytic reaction. The quantum effects in the one-dimensional quantum well of the Pt thin film in Cartesian coordinates is of interest for potential future applications.

CONCLUSION

In this chapter, PEALD was used to deposit an 8-nm-thick Pt nano-thin film using a low-temperature fabrication process. The PEALD-deposited Pt thin films were chemically stable under light illumination and could serve as catalysts under strongly alkaline conditions during the long-term oxidization of ammonium ions. Light illumination with a halogen lamp further enhanced the processing efficiency of ammonium oxidization in an SDR. The experiments with Pt thin films on the reaction disk yielded higher processing efficiencies than that with a bare glass reaction disk. The light absorbed by the Pt thin film can be transformed to compensate the chemical energy in the photocatalytic reaction. The deposited Pt thin film served as a one-dimensional quantum well for the transformation of plasmons to compensate for the chemical energy. The plasmonic photocatalytic enhancement reduced the energy needed in the endothermic oxidization of NH_4^+ believed to be originating from the effect of generated hot electrons on the Pt thin film under light illumination. The energy transformation in metal structures with confinement on the nanoscale enables an interesting physical discussion of quantum electrodynamics.

REFERENCES

[1] Suntola, T. and J. Antson, *Method for producing compound thin films.* 1977: United States patent.

[2] Suntola, T., Atomic layer epitaxy. *Materials Science Reports*, 1989. 4(5): p. 261-312.

[3] Leskelä, M. and M. Ritala, Atomic layer deposition (ALD): from precursors to thin film structures. *Thin Solid Films*, 2002. 409(1): p. 138-146.

[4] Kukli, K., et al., Tailoring the dielectric properties of HfO_2–Ta_2O_5 nanolaminates. *Applied Physics Letters*, 1996. 68(26): p. 3737-3739.

[5] Nishizawa, J.i., H. Abe, and T. Kurabayashi, Molecular Layer Epitaxy. *Journal of The Electrochemical Society*, 1985. 132(5): p. 1197-1200.

[6] Ritala, M., et al., Perfectly Conformal TiN and Al_2O_3 Films Deposited by Atomic Layer Deposition. *Chemical Vapor Deposition*, 1999. 5(1): p. 7-9.

[7] Cianci, E., et al., Atomic layer deposited TiO_2 for implantable brain-chip interfacing devices. *Thin Solid Films*, 2012. 520(14): p. 4745-4748.

[8] Tien, T.-C., et al., Growth mode transition of atomic layer deposited Al_2O_3 on porous TiO_2 electrodes of dye-sensitized solar cells. *Thin Solid Films*, 2012. 520(6): p. 1745-1750.

[9] Trinh, H.D., et al., The influences of surface treatment and gas annealing conditions on the inversion behaviors of the atomic-layer-deposition Al_2O_3-$In_{0.53}Ga_{0.47}As$ metal-oxide-semiconductor capacitor. *Applied Physics Letters,* 2010. 97(4): p. 042903-3.

[10] Vilhunen, S., et al., Atomic layer deposited TiO_2 films in photodegradation of aqueous salicylic acid. *Separation and Purification Technology*, 2009. 66(1): p. 130-134.

[11] Bedair, S.M., et al., Atomic layer epitaxy of III-V binary compounds. *Applied Physics Letters*, 1985. 47(1): p. 51-53.

[12] Kim, H., et al., Robust TaN_x diffusion barrier for Cu-interconnect technology with subnanometer thickness by metal-organic plasma-enhanced atomic layer deposition. *Journal of Applied Physics*, 2005. 98: p. 014308.

[13] Lee, B.H., et al., UV-enhanced atomic layer deposition of ZrO_2 thin films at room temperature. *Thin Solid Films*, 2010. 518(22): p. 6432-6436.

[14] Knoops, H.C.M., et al., Remote Plasma ALD of Platinum and Platinum Oxide Films. *Electrochemical and Solid-State Letters*, 2009. 12(7): p. G34-G36.

[15] Song, Y.S., et al., Optical and electrical properties of transparent conducting gallium-doped ZnO electrodes prepared by atomic layer deposition for application in organic solar cells. *Thin Solid Films*, 2013. 546: p. 271-274.

[16] Ahn, C.H., H. Kim, and H.K. Cho, Deposition of Al doped ZnO layers with various electrical types by atomic layer deposition. *Thin Solid Films*, 2010. 519(2): p. 747-750.

[17] Kim, S., et al., Characteristics of HfO_2 thin films deposited by plasma-enhanced atomic layer deposition using O_2 plasma and N_2O plasma. *Journal of Vacuum Science & Technology B: Microelectronics and Nanometer Structures*, 2006. 24(3): p. 1088-1093.

[18] Xie, Q., et al., Growth Kinetics and Crystallization Behavior of TiO_2 Films Prepared by Plasma Enhanced Atomic Layer Deposition. *Journal of The Electrochemical Society*, 2008. 155(9): p. H688-H692.

[19] Kwon, O.-K., et al., PEALD of a Ruthenium Adhesion Layer for Copper Interconnects. *Journal of The Electrochemical Society*, 2004. 151(12): p. C753-C756.

[20] Kwon, S.-H., et al., Initial Stages of Ruthenium Film Growth in Plasma-Enhanced Atomic Layer Deposition. *Journal of The Electrochemical Society*, 2008. 155(5): p. H296-H300.

[21] Kim, S.-W., et al., Improvement of Copper Diffusion Barrier Properties of Tantalum Nitride Films by Incorporating Ruthenium Using PEALD. *Journal of The Electrochemical Society*, 2008. 155(11): p. H885-H888.

[22] Zhu, Y., K.A. Dunn, and A.E. Kaloyeros, Properties of ultrathin platinum deposited by atomic layer deposition for nanoscale copper-metallization schemes. *Journal of Materials Research*, 2007. 22(05): p. 1292-1298.

[23] Aaltonen, T., et al., Atomic Layer Deposition of Platinum Thin Films. *Chemistry of Materials*, 2003. 15(9): p. 1924-1928.

[24] Jiang, X. and S.F. Bent, Atomic Layer Deposition of Platinum for Solid Oxide Fuel Cells. *ECS Transactions*, 2007. 3(15): p. 249-259.

[25] Grubbs, R.K., et al., Nucleation and growth during the atomic layer deposition of W on Al_2O_3 and Al_2O_3 on W. *Thin Solid Films*, 2004. 467(1-2): p. 16-27.

[26] Li, Z., A. Rahtu, and R.G. Gordon, Atomic Layer Deposition of Ultrathin Copper Metal Films from a Liquid Copper(I) Amidinate Precursor. *Journal of The Electrochemical Society*, 2006. 153(11): p. C787-C794.

[27] Liu, C., et al., Atomic Layer Deposition of Platinum Nanoparticles on Carbon Nanotubes for Application in Proton-Exchange Membrane Fuel Cells. *Small,* 2009. 5(13): p. 1535-1538.

[28] Knoops, H.C.M., et al., Deposition of TiN and TaN by Remote Plasma ALD for Cu and Li Diffusion Barrier Applications. *Journal of The Electrochemical Society*, 2008. 155(12): p. G287-G294.

[29] Liu, B.-H., et al., Platinum thin films with good thermal and chemical stability fabricated by inductively coupled plasma-enhanced atomic layer deposition at low temperatures. *Thin Solid Films*, 2014. 566(0): p. 93-98.

[30] Wu, L. and E. Eisenbraun, Effects of Hydrogen Plasma Treatments on the Atomic Layer Deposition of Copper. *Electrochemical and Solid-State Letters*, 2008. 11(5): p. H107-H110.

[31] Huang, H.J., B.-H. Liu, and J.A. Yeh, Ammonium oxidization at room temperature and plasmonic photocatalytic enhancement. *Catalysis Communications,* 2013. 36(0): p. 16-19.

[32] Huang, H.J. and B.-H. Liu, Plasmonic Energy Transformation in the Photocatalytic Oxidation of Ammonium. *Catalysis Communications*, 2014. 43: p. 136-140.
[33] Huang, H.J., K.-C. Huang, and D.P. Tsai, The Light Absorption Measurement of a Plasmonic Photocatalyst in the Circular Plane Waveguide of a Photocatalytic Dual Light Source Spinning Disk Reactor. *Optical Review*, 2013. 20: p. 236-240.
[34] Lee, D.K., J.S. Cho, and W.L. Yoon, Catalytic wet oxidation of ammonia: Why is N_2 formed preferentially against ? *Chemosphere,* 2005. 61(4): p. 573-578.
[35] Kim, S. and W. Choi, Kinetics and Mechanisms of Photocatalytic Degradation of $(CH_3)_nNH_4^-$ $(0 \leq n \leq 4)$ in TiO_2 Suspension: The Role of OH Radicals. *Environmental Science & Technology*, 2002. 36(9): p. 2019-2025.
[36] Hung, C.-M., J.-C. Lou, and C.-H. Lin, Removal of ammonia solutions used in catalytic wet oxidation processes. *Chemosphere,* 2003. 52(6): p. 989-995.
[37] Huang, L., et al., Removal of Ammonia by OH Radical in Aqueous Phase. *Environmental Science & Technology*, 2008. 42(21): p. 8070-8075.
[38] Zhang, G., D. Yang, and E. Sacher, X-ray Photoelectron Spectroscopic Analysis of Pt Nanoparticles on Highly Oriented Pyrolytic Graphite, Using Symmetric Component Line Shapes. *The Journal of Physical Chemistry C*, 2006. 111(2): p. 565-570.
[39] Hecht, E., Optics. 2002: Addison-Wesley Longman, Incorporated.
[40] Huang, H.J., D.P. Tsai, and A.Q. Liu. A Plasmonic Liquid Waveguide Sensor Using Nanoparticles for Label-Free Measurement Applications in Miniaturized Systems for Chemistry and Life Sciences (μTAS 2008). 2008. San Diego, California, USA.
[41] Lin, Y.-H. and D.P. Tsai, Near-field scanning optical microscopy using a super-resolution cover glass slip. *Opt. Express*, 2012. 20(15): p. 16205-16211.
[42] Liu, A.Q., et al., Label-free detection with micro optical fluidic systems (MOFS): a review. *Analytical and Bioanalytical Chemistry*, 2008. 391(7): p. 2443-2452.
[43] Ma, R.-M., et al., Room-temperature sub-diffraction-limited plasmon laser by total internal reflection. *Nat Mater,* 2011. 10(2): p. 110-113.
[44] Raether, H., Surface Plasmons on Smooth and Rough Surfaces and on Gratings. 1988: Springer.
[45] Ditlbacher, H., et al., Silver Nanowires as Surface Plasmon Resonators. *Physical Review Letters*, 2005. 95(25): p. 257403.
[46] Ebbesen, T.W., et al., Extraordinary optical transmission through sub-wavelength hole arrays. *Nature*, 1998. 391(6668): p. 667-669.
[47] Huang, H.J., et al., Plasmonic optical properties of a single gold nano-rod. *Opt. Express*, 2007. 15(12): p. 7132-7139.
[48] Lin, Y.-H., et al., Near-Field Optical Imaging of a Porous Au Film: Influences of Topographic Artifacts and Surface Plasmons. *Plasmonics,* 2012: p. 1-7.
[49] Moskovits, M., Surface-enhanced spectroscopy. *Reviews of Modern Physics,* 1985. 57(3): p. 783-826.
[50] Nie, S. and S.R. Emory, Probing Single Molecules and Single Nanoparticles by Surface-Enhanced Raman Scattering. *Science,* 1997. 275(5303): p. 1102-1106.
[51] Ono, A., J.-i. Kato, and S. Kawata, Subwavelength Optical Imaging through a Metallic Nanorod Array. *Physical Review Letters*, 2005. 95(26): p. 267407.
[52] Seidel, J., et al., Coupling between surface plasmon modes on metal films. *Physical Review B,* 2004. 69(12): p. 121405.

[53] Sönnichsen, C., et al., Drastic Reduction of Plasmon Damping in Gold Nanorods. *Physical Review Letters*, 2002. 88(7): p. 077402.

[54] Tsai, D.P. and W.C. Lin, Probing the near fields of the super-resolution near-field optical structure. *Applied Physics Letters*, 2000. 77(10): p. 1413-1415.

[55] Lin, C.-T., et al., A simple fabrication process of Pt-TiO_2 hybrid electrode for photo-assisted methanol fuel cells. *Microelectronic Engineering,* 2011. 88(8): p. 2644-2646.

[56] Liu, M., et al., Light-driven nanoscale plasmonic motors. *Nat Nano,* 2010. 5(8): p. 570-573.

[57] Li, Y. and J.N. Armor, Selective NH_3 oxidation to N_2 in a wet stream. *Applied Catalysis B: Environmental*, 1997. 13(2): p. 131-139.

[58] Curtin, T., et al., The catalytic oxidation of ammonia: influence of water and sulfur on selectivity to nitrogen over promoted copper oxide/alumina catalysts. *Catalysis Today*, 2000. 55(1-2): p. 189-195.

[59] Lietti, L., et al., Characterization and reactivity of MoO_3/SiO_2 catalysts in the selective catalytic oxidation of ammonia to N_2. *Catalysis Today,* 2000. 61(1-4): p. 187-195.

[60] Cao, S., et al., Catalytic wet air oxidation of wastewater containing ammonia and phenol over activated carbon supported Pt catalysts. *Catalysis Today*, 2003. 88(1-2): p. 37-47.

[61] Cao, L., J. Yang, and J. Jia, Electro-thermal treatment of high concentration ammonia in water by gaseous oxidation in liquid phase (GOLP). *Chemosphere,* 2010. 80(4): p. 463-468.

[62] Dosta, J., et al., Operation and model description of a sequencing batch reactor treating reject water for biological nitrogen removal via nitrite. *Bioresource Technology*, 2007. 98(11): p. 2065-2075.

[63] Goldstein, R.J., Marine Reef Aquarium Handbook. 1997: Barron's Educational Serious, Inc.

[64] Low, G.K.C., S.R. McEvoy, and R.W. Matthews, Formation of nitrate and ammonium ions in titanium dioxide mediated photocatalytic degradation of organic compounds containing nitrogen atoms. *Environmental Science & Technology*, 1991. 25(3): p. 460-467.

[65] Schmelling, D.C. and K.A. Gray, Photocatalytic transformation and mineralization of 2,4,6-trinitrotoluene (TNT) in TiO2 slurries. *Water Research*, 1995. 29(12): p. 2651-2662.

[66] Yatmaz, H.C., C. Wallis, and C.R. Howarth, The spinning disc reactor – studies on a novel TiO2 photocatalytic reactor. *Chemosphere,* 2001. 42(4): p. 397-403.

[67] Van Gerven, T., et al., A review of intensification of photocatalytic processes. *Chemical Engineering and Processing: Process Intensification*, 2007. 46(9): p. 781-789.

[68] Dionysiou, D.D., et al., Rotating disk photocatalytic reactor: development, characterization, and evaluation for the destruction of organic pollutants in water. *Water Research*, 2000. 34(11): p. 2927-2940.

[69] Chen, Y.L., et al., ZnO nanorod optical disk photocatalytic reactor for photodegradation of methyl orange. *Opt. Express*, 2013. 21(6): p. 7240-7249.

[70] Chang, C.-Y. and N.-L. Wu, Process Analysis on Photocatalyzed Dye Decomposition for Water Treatment with TiO_2-Coated Rotating Disk Reactor. *Industrial & Engineering Chemistry Research*, 2010. 49(23): p. 12173-12179.

[71] Burns, J.R. and R.J.J. Jachuck, Determination of liquid-solid mass transfer coefficients for a spinning disc reactor using a limiting current technique. *International Journal of Heat and Mass Transfer*, 2005. 48(12): p. 2540-2547.

[72] Boodhoo, K.V.K. and R.J. Jachuck, Process intensification: spinning disk reactor for styrene polymerisation. *Applied Thermal Engineering,* 2000. 20(12): p. 1127-1146.

[73] Lin, C.C., C., et al., Photocatalytic degradation of methyl orange by a multi-layer rotating disk reactor. *Environmental Science and Pollution Research,* 2012. 19(9): p. 3743-50.

[74] The Engineering ToolBox. Available from: http://www.engineeringtoolbox.com/ specific-heat-metals-d_152.html.

[75] The refractive index, available from: http://refractiveindex.info/. Pt was from http://refractiveindex.info/?shelf=main&book=Pt&page=Rakic. Water was http://refractiveindex.info/?shelf=3d&book=liquids& page=water. Glass was from http://refractiveindex.info/?shelf =glass&book=BK7&page=SCHOTT.

[76] 76 Abe, R., et al., Pristine Simple Oxides as Visible Light Driven Photocatalysts: Highly Efficient Decomposition of Organic Compounds over Platinum-Loaded Tungsten Oxide. *Journal of the American Chemical Society,* 2008. 130(25): p. 7780-7781.

[77] Awazu, K., et al., A Plasmonic Photocatalyst Consisting of Silver Nanoparticles Embedded in Titanium Dioxide. *Journal of the American Chemical Society*, 2008. 130(5): p. 1676-1680.

[78] Chen, J.-J., et al., Plasmonic Photocatalyst for H2 Evolution in Photocatalytic Water Splitting. *The Journal of Physical Chemistry C,* 2010. 115(1): p. 210-216.

[79] Kuai, L., et al., Facile Subsequently Light-Induced Route to Highly Efficient and Stable Sunlight-Driven Ag–AgBr Plasmonic Photocatalyst. *Langmuir,* 2010. 26(24): p. 18723-18727.

[80] Mostafa A, E.-S., Plasmonic photochemistry and photon confinement to the nanoscale. *Journal of Photochemistry and Photobiology A: Chemistry,* 2011. 221(2–3): p. 138-142.

[81] Tian, Y. and T. Tatsuma, Mechanisms and Applications of Plasmon-Induced Charge Separation at TiO2 Films Loaded with Gold Nanoparticles. *Journal of the American Chemical Society*, 2005. 127(20): p. 7632-7637.

[82] Yu, J., G. Dai, and B. Huang, Fabrication and Characterization of Visible-Light-Driven Plasmonic Photocatalyst Ag/AgCl/TiO2 Nanotube Arrays. *The Journal of Physical Chemistry* C, 2009. 113(37): p. 16394-16401.

[83] Avanesian, T. and P. Christopher, Adsorbate Specificity in Hot Electron Driven Photochemistry on Catalytic Metal Surfaces. *The Journal of Physical Chemistry C*, 2014. 118(48): p. 28017-28031.

[84] Govorov, A.O., et al., Photogeneration of hot plasmonic electrons with metal nanocrystals: Quantum description and potential applications. *Nano Today*, 2014. 9(1): p. 85-101.

[85] Govorov, A.O., H. Zhang, and Y.K. Gun'ko, Theory of Photoinjection of Hot Plasmonic Carriers from Metal Nanostructures into Semiconductors and Surface Molecules. *The Journal of Physical Chemistry C*, 2013. 117(32): p. 16616-16631.

[86] Adleman, J.R., et al., Heterogenous Catalysis Mediated by Plasmon Heating. *Nano Letters, 2009*. 9(12): p. 4417-4423.

[87] Clavero, C., Plasmon-induced hot-electron generation at nanoparticle/metal-oxide interfaces for photovoltaic and photocatalytic devices. *Nat Photon*, 2014. 8(2): p. 95-103.

[88] Du, L., et al., Ultrafast plasmon induced electron injection mechanism in gold–TiO_2 nanoparticle system. *Journal of Photochemistry and Photobiology C: Photochemistry Reviews,* 2013. 15(0): p. 21-30.

[89] Link, S. and M.A. El-Sayed, Spectral Properties and Relaxation Dynamics of Surface Plasmon Electronic Oscillations in Gold and Silver Nanodots and Nanorods. *The Journal of Physical Chemistry B,* 1999. 103(40): p. 8410-8426.

[90] Manjavacas, A., et al., Plasmon-Induced Hot Carriers in Metallic Nanoparticles. *ACS Nano,* 2014. 8(8): p. 7630-7638.

[91] Mukherjee, S., et al., Hot Electrons Do the Impossible: Plasmon-Induced Dissociation of H2 on Au. *Nano Letters*, 2013. 13(1): p. 240-247.

INDEX

C

D

E

F

G

H

I

J

K

L

M

N

O

P

Q

R

S

T

U

V

W

X

Y

Z